AF779999

Sustainable Energy Transition

A Comprehensive Guide to Modelling and Optimizing Energy Systems

Editors

Chandra Prakash Garg

Department of Operations Management & Quantitative Techniques
Indian Institute of Management Rohtak, Haryana, India

Atul Rawat

Department of Oil & Gas Management
School of Business
University of Petroleum & Energy Studies, Dehradun, India

Priyank Sinha

Department of Industrial Engineering and Operations Research
Indian Institute of Technology, Bombay, India

CRC Press
Taylor & Francis Group
Boca Raton London New York

CRC Press is an imprint of the
Taylor & Francis Group, an **informa** business
A SCIENCE PUBLISHERS BOOK

Cover credit: Cover image created by the editors using copyright-free resources.

First edition published 2027
by CRC Press
2385 NW Executive Center Drive, Suite 320, Boca Raton FL 33431

and by CRC Press
4 Park Square, Milton Park, Abingdon, Oxon, OX14 4RN

© 2027 Chandra Prakash Garg, Atul Rawat and Priyank Sinha

CRC Press is an imprint of Taylor & Francis Group, LLC

Reasonable efforts have been made to publish reliable data and information, but the author and publisher cannot assume responsibility for the validity of all materials or the consequences of their use. The authors and publishers have attempted to trace the copyright holders of all material reproduced in this publication and apologize to copyright holders if permission to publish in this form has not been obtained. If any copyright material has not been acknowledged please write and let us know so we may rectify in any future reprint.

Except as permitted under U.S. Copyright Law, no part of this book may be reprinted, reproduced, transmitted, or utilized in any form by any electronic, mechanical, or other means, now known or hereafter invented, including photocopying, microfilming, and recording, or in any information storage or retrieval system, without written permission from the publishers.

For permission to photocopy or use material electronically from this work, access www.copyright.com or contact the Copyright Clearance Center, Inc. (CCC), 222 Rosewood Drive, Danvers, MA 01923, 978-750-8400. For works that are not available on CCC please contact mpkbookspermissions@tandf.co.uk

For Product Safety Concerns and Information please contact our EU representative GPSR@taylorandfrancis.com. Taylor & Francis Verlag GmbH, Kaufingerstraße 24, 80331 München, Germany.

Trademark notice: Product or corporate names may be trademarks or registered trademarks and are used only for identification and explanation without intent to infringe.

Library of Congress Cataloging-in-Publication Data (applied for)

ISBN: 978-1-032-74163-5 (hbk)
ISBN: 978-1-032-75725-4 (pbk)
ISBN: 978-1-003-47538-5 (ebk)

DOI: 10.1201/9781003475385

Typeset in Times New Roman
by Prime Publishing Services

Preface

The 21st century has brought unprecedented attention to climate change, environmental degradation, and the urgent need for a global shift in how energy is produced, consumed, and governed. At the heart of this transition lies a transformative opportunity: to reshape the energy systems that power our economies and societies into sustainable, resilient, and inclusive frameworks. This book, *Sustainable Energy Transition: A Comprehensive Guide to Modeling and Optimizing Energy Systems*, is a response to this imperative. It seeks to provide a holistic and research-backed exploration of how nations—especially emerging economies like India—can navigate the complex pathway toward a cleaner and more equitable energy future.

India, as one of the fastest-growing economies and the third-largest energy consumer globally, stands at a pivotal junction in the global energy landscape. With 17% of the world's population and less than 4% of global carbon emissions, the country has taken significant strides in aligning with its climate commitments under the Paris Agreement. It has actively championed clean energy policies through its G20 leadership and COP28 representation, emerging as a powerful voice of the Global South. Yet, the country also grapples with a host of structural, financial, and technological challenges that impede its journey toward a fully sustainable energy future.

The book opens with an analysis of these dynamics in Chapter 1, which critically examines the state of renewable energy deployment in India. It highlights the country's commendable growth in installed renewable capacity—reaching over 209 GW by December 2024—and unpacks the barriers stifling further progress. From financially distressed distribution companies (DISCOMs) to reliance on imported critical minerals and policy inconsistencies, the chapter uses the Fuzzy Analytic Hierarchy Process (FAHP) to systematically identify and prioritize these challenges. This approach allows stakeholders to understand not only the severity of the issues but also the interconnections that make energy transition a multi-layered governance challenge.

Chapter 2 builds on this foundation by focusing on the indispensable role of energy planning in achieving the Paris Agreement's ambitious targets. The research combines literature review, regional case studies from Tamil Nadu and Pune, and media analysis to underscore how decentralized and well-integrated planning strategies can lead to robust, participatory, and socially beneficial renewable energy ecosystems. The chapter argues that successful energy transitions rely on both infrastructure and people—engaging local communities, creating employment, and ensuring health and welfare benefits. It also emphasizes how decentralization, if aligned with strong governance, can foster greater resilience and equity in energy access.

Taking a global view, Chapter 3 examines the international landscape of renewable energy transitions. It reflects on how countries worldwide are moving away from fossil fuels and investing in greener alternatives, often facing similar obstacles but also learning from shared experiences. Through policy analysis, case study reviews, and the lens of the Sustainable Development Goals (SDGs), this chapter explores how technological innovation, inclusive governance, and social participation must go hand in hand. Special emphasis is placed on SDG 12 (Responsible Consumption and Production) and SDG 17 (Partnerships for the Goals), reinforcing the notion that energy transition is not only a national concern but also a global collaboration.

Shifting the focus back to India, Chapter 4 delves deeper into decarbonization strategies, especially highlighting the underexplored area of demand-side energy management. Traditionally, energy policy has emphasized the supply side—adding more renewable generation capacity. However, this chapter makes a compelling case for greater attention to how energy is consumed, advocating for electrification of systems and behavior changes as cost-effective and sustainable solutions. Through a detailed review of energy literature, it evaluates Carbon Capture, Storage, and Utilization (CCSU) technologies and concludes that while promising in theory, CCSU may not be practical for developing countries due to high costs and infrastructure demands. Instead, it recommends a strategic focus on renewable energy expansion and demand-side optimization as more viable pathways for India's energy decarbonization.

As energy demand continues to grow—especially in urbanized regions—the importance of choosing the right optimization tools and strategies becomes paramount. Chapter 5 introduces Grey Multiple Criteria Decision Making (Grey MCDM) methods as a powerful framework for making informed choices under conditions of uncertainty and incomplete information. Through hypothetical case studies, the chapter evaluates various energy solutions such as rooftop solar, microgrids, energy storage, and traditional grid upgrades. It analyzes these options based on criteria including cost, efficiency, scalability, and emissions impact. The result is a practical decision-making toolkit that policymakers and planners can use to develop sustainable and resilient energy infrastructure tailored to local needs.

The energy transition also brings unique challenges to high-emission sectors such as aviation, which is the focus of Chapter 6. The aviation industry's roadmap to net-zero emissions by 2050 requires more than just incremental improvements. This chapter examines the techno-economic feasibility of various clean fuel alternatives, including Sustainable Aviation Fuels (SAFs), liquid hydrogen, and battery-electric propulsion. It presents a comparative analysis of these options against conventional jet fuels, highlighting their environmental benefits, technical barriers, and market readiness. SAFs, while immediately deployable and capable of reducing emissions by up to 80%, are constrained by supply limitations and high costs. Meanwhile, hydrogen and battery technologies are promising but face challenges in energy density and infrastructure requirements. The chapter concludes with a call for a socio-technical approach—one that integrates technological innovation with economic incentives, policy support, and stakeholder engagement—to achieve a sustainable aviation future.

Chapter 7 turns to the issue of energy security, which is increasingly relevant amid geopolitical tensions, climate risks, and the volatility of global energy markets. Using the Decision-Making Trial and Evaluation Laboratory (DEMATEL) methodology, the chapter identifies and ranks key barriers to achieving energy security in emerging economies. It uncovers causal relationships between factors such as policy gaps, financial constraints, technological limitations, and social resistance. The findings provide actionable insights into where interventions can be most effective, offering a structured pathway for governments, planners, and stakeholders to enhance their national energy security frameworks.

The built environment also plays a crucial role in the energy landscape, particularly as buildings account for a significant portion of energy consumption and emissions. Chapter 8 explores how Building Information Modeling (BIM) integrated with Building Energy Modeling (BEM) can revolutionize the design and management of energy-efficient structures. It highlights how BIM-BEM integration overcomes traditional limitations such as long model development times and high costs by enabling real-time data collaboration among architects, engineers, and designers. The chapter delves into practical applications during early-stage planning, focusing on HVAC systems, thermal properties, and sustainable material use. It also provides practical guidelines for professionals to implement BIM-BEM methods effectively, showcasing how digital innovation can support the construction of greener, smarter buildings.

Finally, Chapter 9 addresses the manufacturing sector's role in supporting the circular economy and achieving net-zero targets through Industry 4.0 technologies. The study surveys over 550 Indian manufacturing firms to assess the adoption and impact of technologies like Cyber-Physical Systems (CPS), the Internet of Things (IoT), and Big Data on circular practices such as resource optimization and emissions reduction. Findings reveal a strong positive correlation between CPS/IoT implementation and improved environmental, economic, and social performance. Interestingly, Big Data showed limited influence, indicating that not all technologies have equal impact. The chapter provides a roadmap for industries to align their operations with the Sustainable Development Goals, particularly those related to responsible production and climate action.

Collectively, the chapters in this book reflect the multidisciplinary and interconnected nature of sustainable energy transitions. From national policy and technological innovation to community engagement and sector-specific strategies, this book presents a comprehensive, data-driven, and practical framework for reimagining how energy systems can evolve to meet the dual challenges of climate change and development.

We hope this book serves as both a resource and a catalyst—for researchers exploring cutting-edge energy models, for policymakers crafting effective strategies, for industry leaders implementing real-world solutions, and for students building the knowledge base for tomorrow's energy landscape. Above all, we aim to inspire a deeper commitment to sustainability, resilience, and equity in the global journey toward a clean energy future.

Chandra Prakash Garg
Atul Rawat
Priyank Sinha

Contents

Chapter 1

Analyzing Barriers to Sustainable Energy Transition in India Using the Fuzzy AHP Approach

Archana Sharma,[1,*] *Vishal Kashav*[2] and *Jasmine Siu Lee Lam*[3]

1. Introduction

Today, the urgency of sustainability with regard to energy has emerged as a global imperative in light of accelerating environmental degradation, depletion of fossil resources, and an ever-looming threat of climate change. Countries and economies worldwide and more directly, emerging markets like India are now taking centre stage in transitioning from fossil fuel-based traditional systems to renewable and cleaner energy alternatives. India is quite an important player in the global energy transition, considering its large population and economy growing extremely rapidly (Dey et al. 2022). Still, the immense potential of renewable energy sources in the form of solar, wind, hydro, and bio-energy has a path to achieving a sustainable energy future in India that is riddled with many barriers and challenges. Understanding these barriers and addressing them is crucial for successful implementation and long-term sustainability. Energy transition is basically a change in technology,

[1] Department of Applied sciences & Humanities, KIET Group of Institutions, Delhi-NCR, Ghaziabad, U.P., India-201206.
[2] Department of Global Supply Chains & Mobility, UPES, Dehradun, India-248007.
[3] Department of Technology, Management and Economics Operations and Supply Chain Management, Technical University of Denmark.
Emails: Vishal.Kashav@ddn.upes.ac.in, jasmlam@dtu.dk
* Corresponding author: jmdarchanasharma@gmail.com

DOI: 10.1201/9781003475385-1

but it is multifaceted, as it incorporates different dimensions of economics, social, political, and environment. Although India has aimed to create cumulative installed capacity from non-fossil fuel-based energy resources at 50% by 2030, it has also committed internationally to reducing the emission intensity of its GDP by 45% by 2030, based on the levels in the base year of 2005. High penetrations of variable and intermittent renewable energies in the energy mix raise very serious issues with respect to grid stability and sustainable power supply. Renewable energy sources are problematic since they are time, climate, season, or geography sensitive. Regulatory, financial, technological, infrastructural, and community challenges impose greatly on the widespread adoption of sustainable energy solutions (Olabi et al. 2023).

The energy demand in India grows fast as the economy expands through urbanization and a rising population. Currently, it is ranked third globally in terms of energy consumption, behind China and the United States. Traditionally, coal has been the bedrock of Indian energy, accounting for more than 70% of electricity generation in the country. High consumption of coal and other fossil fuels leads not only to more accumulation in the air through pollution but also to significant greenhouse gas emissions. Being a part of the Paris Agreement, India should reduce its carbon emission intensity and enhance the contribution of renewable energy sources to its energy supply mix. India has enormous potential in renewable energy, with rich solar, wind, hydro, and biomass resources spread across the country (Chateau et al. 2023). In fact, the government has taken certain efforts toward renewable energy by establishing the International Solar Alliance, launching the National Solar Mission, and providing numerous subsidies and incentives for renewable energy projects. However, renewable energy is still used in a negligible proportion of the total energy consumed by India. Integration of renewable energy sources into the existing energy grid is still a significant challenge to overcome, both in structure and policy. This paper aims to discuss and model the various barriers impeding the sustainable energy transition in India with the ultimate aim of providing a comprehensive framework for policymakers, industry stakeholders, and researchers to systematically address the barriers (Pouresmaieli et al. 2023).

Renewable energy thereby becomes an essential supply chain to meet the huge demands of renewable energy for the never-before-witnessed numbers in populations in developing regions that are still without access to clean sources of energy (Nastasi et al. 2022). The interaction between rising populations and elevated energy demands in these regions inspires a need for more renewable, sustainable, affordable, and environmentally resource-friendly energy systems. In particular, India, with all its vastness, is very well endowed with renewable energy resources spread across the country, having an overall potential in excess of its current total energy consumption. Considering the challenge of global warming and climate change, India should hasten the implementation of strategies targeted at increasing the share of renewable energy in its overall energy mix (Pillai and Banerjee 2009).

Such a shift is not only conducive to environmental sustainability, but it is also instrumental to economic growth because renewable energy can provide an opportunity that is attractive for investment with scope for long-term economic development in India (Moustakas et al. 2020). Although renewable energy sources are technologically achievable and economically feasible for many applications, they still represent a negligible proportion of the potential available, mainly because of the multiple barriers that face the deployment of green energy technologies, which differ from country to country. Since the acceptance and deployment of renewable or sustainable energy technologies in the Indian context remain critically hindered, it becomes essential that major barriers be identified and addressed. The primary objectives of this paper are to:

- To recognize the main hindrances to adopting renewable or sustainable energy technologies in India.
- To evaluate and rank these barriers in their relative impact on the adoption of renewable or sustainable energy transition in India.

The identification of the barriers is done by systematic literature review and inputs from industrial experts. Then the finalization of the barriers is by a decision-making group. After the identification, these barriers are assessed and ranked by the fuzzy analytic hierarchy process method. The FAHP-based multi-criteria decision-making framework provides a method to effectively handle uncertainty and imprecision in decision-making.

The rest of this chapter is organized as follows. Section 2 presents a detailed literature review of SETA that would help to understand the topic and identify barriers. Section 3 defines barriers to sustainable energy transition adoption to articulate the research problem. The methodology, or more specifically the FAHP process, is elaborated in Section 4 and presents the ranking of the identified barriers. Section 5 reveals the results and discussion. Finally, Section 6 summarizes the chapter as the conclusion.

2. Literature Review

The existing literature on the adoption of sustainable energy technologies in India has identified several key barriers. A study on the barriers to sustainable consumer behavior conducted by many pioneering researchers in the context of the Indian market revealed that the critical barriers can be categorized into a few important constraints. The literature of this study concluded that Fuzzy AHP is very effective in identifying and ranking the barriers to the sustainable energy transition (Mahtani and Garg 2018). Financial constraints remain a consistent issue, indicating that innovative financing schemes and policy support are needed for India's energy transition. Singh and Jain (2022) investigated multiple scenarios pertaining to energy transition in India, emphasizing the integration of renewable energy sources. They pinpointed financial obstacles as a significant hindrance to the expansion of green energy infrastructure and proposed that decentralized energy systems might alleviate these challenges. Wang et al. (2023) analyzed innovation in green energy in China by comparative analytical methods. Their findings are relevant to India because they identified key strategies, including public-private partnerships and subsidies, that can be used to overcome financial constraints in the deployment of renewable energy. Another major hurdle is technology and innovation, which is also hampering the adoption of sustainable energy transition in India. Bamati et al. (2020) analyzed how technological factors affect renewable energy production in the developed and developing world. They proposed a generalized least squares methodology, hypothesizing that high-technology exports remarkably drive renewable energy in developed countries but not in developing ones. Akusta and Cergibozan (2024) discussed the challenges of renewable investment in Turkey with Fuzzy AHP and determined technological inefficiency and outdated systems as major problems. Li et al. (2024) improved the Fuzzy Analytic Hierarchy Process methodology with cumulative prospect theory for the evaluation of renewable energy tracks in Malaysia. The results showed that innovation-related barriers were dominated by underfunding for research and development purposes, as well as a lack of technical skills.

Furthermore, Prakash et al. (2018) focused on the importance of government incentives, consumer characteristics, and infrastructure development as crucial factors for wider EV penetration in India. Many authors have observed key obstacles such as lack of awareness, insufficient infrastructure, and regulatory challenges to sustain the energy transition. The FAHP provides organization-specific actionable insights through its ranking of barriers. The study underscores the importance of improved education together with infrastructure investments and supportive government policies for advancing sustainability practices (Kumar and Garg 2017). The research helps advance knowledge about the difficulties of implementing effective sustainability strategies. Souvik and Ganguly (2017) conducted a study on renewable energy alternatives that support sustainable development by addressing climate change and energy requirements while examining government-backed policy obstacles and possibilities for sustainable growth.

Developed countries often focus on economic and environmental factors as key drivers for successful sustainable energy transitions. However, India faces unique challenges, such as a lack of community support and difficulties in acquiring land for renewable energy projects. Yaqoot et al. (2016) identified several obstacles to decentralized renewable energy systems in India, including outdated technology, a shortage of skilled workers, high costs, and limited access to credit. Their research highlights the importance of creating long-term policies, regulatory frameworks, and financial incentives to encourage the adoption of renewable energy. Additionally, they emphasize the need to engage communities, raise public awareness, and establish institutions to support these systems. Similarly, Indrajayanthan et al. (2022) studied the potential for clean energy transitions in India's key power-producing regions. They found that financial constraints are the biggest challenge to implementing renewable energy technologies, followed by regulatory issues. Their research stresses the importance of aligning state-level policies and offering financial incentives to overcome these barriers. To address these challenges, advanced tools like the Fuzzy Analytical Hierarchy Process (Fuzzy AHP) were used to evaluate the obstacles and prioritize solutions. The study concluded that financial limitations, along with regulatory hurdles, are the main factors slowing down the large-scale adoption of renewable energy in India.

In developing countries like India, the transition to sustainable energy is complex, involving financial, technological, and policy-related challenges. Researchers have used advanced methods, such as Fuzzy AHP, to analyze these issues and identify practical solutions. For example, Mishra (2023) introduced a new approach to regulatory governance by using mixed methods, including document and thematic analysis, to identify key factors that could improve industry performance. Globally, the energy crisis has also influenced energy consumption patterns. Einolander et al. (2024) studied how Finnish households responded to rising electricity prices during the global energy crisis. Using smart meter data and cluster analysis, they found that about one-third of households, particularly those using nighttime water heating, oil heating, or ground-source heating, showed significant changes in their energy usage. The study demonstrated that real-time pricing contracts effectively encourage energy-saving behaviors and help shape energy policies. Similarly, other authors have also confirmed these findings, showing that real-time pricing contracts promote energy-efficient practices among consumers. In South Asia, Muhammad (2024) explored energy challenges in countries like India, Pakistan, Bangladesh, and Sri Lanka. The study highlighted issues such as limited energy access, unreliable grids, and the need to expand renewable energy. It also incorporated insights from stakeholders to identify opportunities for a sustainable energy transition. These findings, along with the key barriers, are summarized in Table 1 below.

Table 1. Identified Barriers of Sustainable Energy Transition Adaption (SETA).

Code	**Barriers**	**References**
B1	Financial Constraints	Wang et al. (2023), Saraswat et al. (2021)
B2	Lack of Technological Adaption and Innovation	Akusta et al. (2024), Li et al. (2024)
B3	Policy and Governance Issues	Sen and Ganguly (2017), Mishra (2023)
B4	Community Acceptances and Land Acquisition Issues	Prakash et al. (2018)
B5	Poor Environmental Regulations and Rules	Yaqoot et al. (2016)
B6	Poor Supply Chain Network	Abhyankar et al. (2023)
B7	Fragmented Market and Price Sensitive Customers	Einolander et al. (2024)

3. Research Methodology

This study introduces a Fuzzy Analytic Hierarchy Process (FAHP) approach to evaluate and prioritize the barriers to adopting sustainable energy transitions (SET). The framework, shown in Figure 1, provides a structured way to identify and rank these barriers, addressing the complexity of the process. It starts with forming a decision-making group that identifies and validates the barriers by reviewing existing research and gathering insights from experts in the field. This stage involved refining inputs through ongoing discussions with experts via email, phone calls, and other communication methods. Once the group reached a consensus, the FAHP

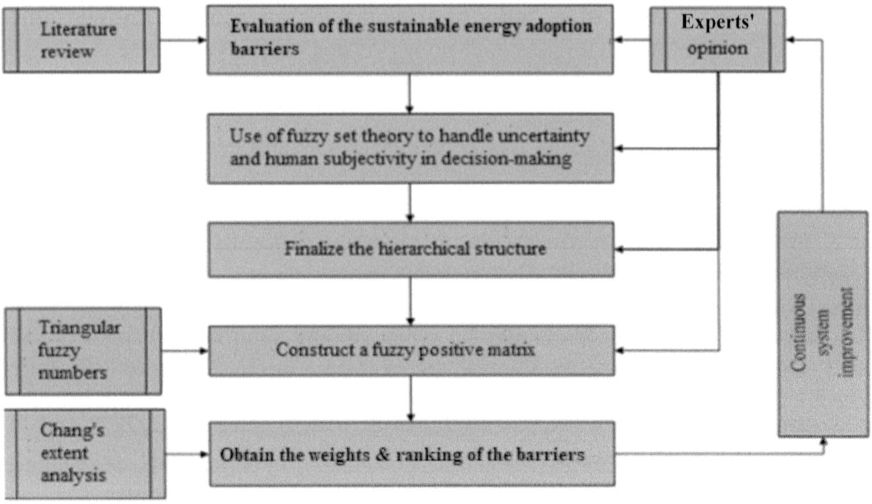

Figure 1. Flowchart of research methodology. Source: Authors' composition.

methodology was used to assess the importance of each barrier. FAHP was chosen because it is particularly effective in handling uncertainty and ambiguity, which are common in decision-making processes involving subjective judgments. Unlike traditional AHP, FAHP uses fuzzy scales instead of fixed numerical values, allowing for a more flexible and detailed evaluation of expert opinions. By incorporating triangular fuzzy numbers, the method captures a wider range of perspectives while ensuring consistency and rationality. These fuzzy numbers also reflect the confidence levels of decision-makers, creating a balanced and reliable way to combine expert input. Involving decision-makers at every step of the process ensures transparency and reduces issues like bias or inconsistency.

This approach offers a practical and adaptable system for evaluating barriers to sustainable energy transitions. The FAHP method simplifies complex decision-making while providing meaningful insights for both researchers and practitioners. It emphasizes accuracy, transparency, and flexibility, making it a valuable tool for tackling strategic challenges in uncertain situations. The study also provides a clear, step-by-step guide to implementing FAHP, making it easier for future researchers and professionals to apply this method in similar scenarios.

3.1 Phase II FAHP Method

The Fuzzy Analytic Hierarchy Process (FAHP) is a popular method for decision-making that combines both quantitative and qualitative factors into a single, structured framework. It is based on linear programming principles and is particularly useful for solving complex problems that involve multiple criteria. Traditional AHP is widely used because it is simple, requires fewer pairwise comparisons, and effectively determines the relative importance of different criteria. This makes it a preferred choice over other decision-making methods like Best-Worst Method (BWM), Multi-Attribute Utility Theory (MAUT), VIKOR, PROMETHEE, and DEMATEL. Unlike the Analytic Network Process (ANP), which deals with complex interdependencies, AHP focuses on straightforward linear relationships, making it easier to evaluate and compare decision variables. Despite its advantages, traditional AHP has some limitations. It relies heavily on subjective judgments and fixed measurement scales, which can sometimes lead to inconsistent or unstable results. To address these issues, fuzzy logic has been integrated with AHP, creating a more reliable and flexible approach. Fuzzy AHP improves the decision-making process by accounting for the uncertainty and vagueness that often come with human judgment. It uses triangular fuzzy numbers (TFNs), as shown in Table 2, to better represent decision-makers' preferences and provide a more accurate evaluation. By incorporating fuzzy logic, FAHP overcomes the challenges of traditional AHP, such as bias and rigid scaling, making it a more dependable tool for decision-making. Its ability to handle uncertainty while remaining easy to use has made it increasingly popular in various fields. Researchers and practitioners value FAHP for its improved accuracy and its ability to provide clear, actionable insights in complex decision-making scenarios. Its combination of flexibility, precision, and practicality has established FAHP as a trusted method for evaluating multiple criteria in a wide range of applications.

The FAHP methodology is applied using Chang's extent analysis approach. This approach employs TFNs to evaluate the fuzzy synthetic extent values for each criterion. The process of implementing Chang's extent analysis can be broken down into the following steps:

Step 1: The fuzzy synthetic extent value is computed for each criterion. This step involves aggregating the fuzzy assessments of decision-makers across all relevant criteria using fuzzy arithmetic operations.

Table 2. Evaluation scale.

Linguistics representation	Dispensed TFN
EI Equally Important	(1, 1, 1)
VL-Very Less Important	(1, 2, 3)
LI-Low Important	(2, 3, 4)
AI-Average Important	(3, 4, 5)
HI-Higher Important	(4, 5, 6)
VHI-Very High Important	(5, 6, 7)
ABI-Absolutely Important	(7, 8, 9)

Source: Authors' composition

Steps 2 and 3: The degree of possibility is determined by comparing the fuzzy synthetic extent values of a given criterion against those of all other criteria. This step establishes the relative priority or importance of each criterion. The relative criterion weights are computed using the degree of possibility, represented as:

$$W' = [d'[A_1], d'[A_2], \ldots\ldots, d'[A_m]]^T \tag{1.1}$$

Step 4: Normalization procedure is employed to derive the normalized weights of the parameters. This step ensures that the weights are scaled appropriately to reflect their relative importance within the decision-making framework. The normalized weights are expressed as:

$$W = [d[A_1], d[A_2], \ldots\ldots, d[A_m]]^T \tag{1.2}$$

3.2 Calculate the Priority Weights

The decision-making team conducted a comparative evaluation of the seven identified SET barriers by assigning fuzzy ratings, as outlined in Table 3, based on the fuzzy scoring framework described in Table 2. Subsequently, the FAHP, as previously detailed, was utilized to determine the weighted importance of each barrier. The calculated weightage scores for the SET barriers are presented in Table 3, showcasing the relative significance of each factor in the overall decision-making framework.

Table 3. Comparison matrix of the SETA barriers.

	B1	B2	B3	B4	B5	B6	B7
B1	(1, 1, 1)	(0.25, 0.33, 0.5)	(3, 4, 5)	(0.33, 0.5, 1)	(1, 2, 3)	(3, 4, 5)	(0.33, 0.5, 1)
B2	(2, 3, 4)	(1, 1, 1)	(0.33, 0.5, 1)	(0.33, 0.5, 1)	(0.33, 0.5, 1)	(3, 4, 5)	(0.25, 0.33, 0.5)
B3	(0.2, 0.25,0.33)	(2, 3, 4)	(1, 1, 1)	(3, 4, 5)	(0.25, 0.33, 0.5)	(3, 4, 5)	(3, 4, 5)
B4	(1, 2, 3)	(1, 2, 3)	(0.2, 0.25,0.33)	(1, 1, 1)	(1, 2, 3)	(0.2, 0.25,0.33)	(0.25, 0.33, 0.5)
B5	(0.33, 0.5, 1)	(1, 2, 3)	(2, 3, 4)	(0.33, 0.5, 1)	(1, 1, 1)	(1, 2, 3)	(0.25, 0.33, 0.5)
B6	(0.2, 0.25,0.33)	(0.2, 0.25,0.33)	(0.2, 0.25,0.33)	(3, 4, 5)	(0.33, 0.5, 1)	(1, 1, 1)	(0.33, 0.5, 1)
B7	(1, 2, 3)	(2, 3, 4)	(0.2, 0.25,0.33)	(2, 3, 4)	(2, 3, 4)	(1, 2, 3)	(1, 1, 1)

Source: Fuzzy AHP analysis by authors.

Table 4. The weightage score and ranking of the SETA barriers.

Barriers	Weightage score	Priority list
B1	0.1697	3
B2	0.1318	4
B3	0.2196	1
B4	0.0959	6
B5	0.1280	5
B6	0.0598	7
B7	0.1950	2

Source: Fuzzy AHP analysis by authors

4. Results and Discussions

The ranking of barriers is determined based on their final weights. After applying the FAHP, the sequence of barriers is as follows: B3 > B7 > B1 > B2 > B5 > B4 > B6; according to the weights arranged in descending order (refer to Table 4). The FAHP analysis highlights that B3, i.e., Policy and Governance Issues, holds the highest rank with a weightage value of 0.2196. This underscores the need for Indian agencies to intervene in the sustainable energy transition, as regulation is a critical driver for implementing policy initiatives by the government. In India, regulatory uncertainty and policy fragmentation are some of the major hurdles ahead in its pursuit of successfully transitioning toward renewable energy. Whereas the national government has ambitiously set targets for renewable energy adoption, the real gap between national and state-level policies is gaping. The decentralized nature of energy governance in India gives much authority to individual states to determine their energy policies, often leaving a mess of inconsistent regulatory frameworks across the nation. This inconsistency creates challenges for renewable energy developers, who must navigate varying policies, regulations, and approval processes in different states. The absence of clear and long-term policy certainty thus undermines investor confidence in renewable energy projects. Such policy changes, if frequent, such as the withdrawal or modification of incentives and subsidies for renewable energy, consequently create an uncertain investment climate. Investors, therefore, are not interested in committing to large-scale renewable energy projects in the absence of proper and stable, long-standing policies.

The second most significant barrier, B7 (0.1950), is fragmented market and price-sensitive customers. The renewable energy market in India is highly diversified with various stakeholders, regional disparities, and varied state policy frameworks that do not favor large-scale adoption. This decentralized energy governance leads to regulatory and incentive disparity and makes it hard for renewable energy developers to enter markets. For example, whereas some states offer very good incentives for both solar and wind energy, others are not as encouraging, and that has led to uneven growth. Highly price-sensitive energy consumers in India have proven to be a challenge to renewable energy uptake. Much as solar and wind energy are becoming cheaper, many consumers, especially those in rural settings and

Chapter 2

Importance of Energy Planning and Transition

Romi Sainy[1,*] and *Manoj Kumar Kamila*[2]

1. Introduction

Global warming presents a significant and urgent threat to our planet. Immediate action is required to address this crisis. The Intergovernmental Panel on Climate Change (IPCC), a body of climate experts, emphasizes the necessity of rapidly reducing greenhouse gas emissions to mitigate excessive global temperature increases. In line with this, over 130 countries have committed to achieving net-zero greenhouse gas emissions by 2050, a target established under the Paris Agreement. Achieving these objectives mandates a cessation of reliance on coal, oil, and gas, which are primary contributors to environmental pollution. Consequently, a transition to clean energy sources, such as solar and wind power, is imperative. Currently, over 80% of global energy is derived from sources that are detrimental to the environment. Changing to clean energy is not just good for the Earth but also for our health and money. It's not easy to make this change, but it is essential. We need good planning to make this work. This means thinking about how it will affect people's jobs and lives and how to set up the new energy systems. If we do this right, we can have a healthier world. Integrated energy planning thus becomes relatively straightforward and relevant. Good energy planning largely depends on comprehending the complex interactions among broader socioeconomic frameworks and energy systems. It offers a strategic framework that helps countries and people evaluate trade-offs, simulate prospective energy futures, appraise their present resources, and build solid and sensible routes to attain these

[1] Professor – Marketing, Jaipuria Institute of Management, Indore, India.
[2] Research Scholar, Jaipuria Institute of Management, Indore, India.
Email: manojkumar.kamila.fpm22i@jaipuria.ac.in
* Corresponding author: romi.sainy@jaipuria.ac.in

DOI: 10.1201/9781003475385-2

energy transitions. Energy influences public health, industry, employment, and economic growth. Thus, planning must be comprehensive and inclusive, embracing governments, companies, communities, and people, among other spheres (Barton 2009). By emphasizing the necessity of the energy transition are the upcoming targets for world carbon emissions. National governments are increasingly crucial in coordinating policies across departments to maximize decarbonization efforts, merging sectoral strategies for power, transportation, buildings, and industry with more general economic policies such as carbon pricing and fossil fuel subsidy reductions. The latest IPCC estimates indicate that global emissions must drop by 45% by 2030 from 2010 levels and reach net zero by 2050 to save a climate path limiting warming to 1.5°C (Shahzad et al. 2024).

Along with strong financial and political support, this ambitious target demands not only fast implementation of present renewable technologies but also significant developments in new technologies, together with the management of the complex social effects accompanying such fundamental transformations in energy systems. These changes involve how land is used, the social and economic aspects of energy access, and shifts in job markets due to the decline of fossil fuel industries and the rise of new energy technologies. Strong energy planning helps us transition to renewable energy systems by fighting climate change while benefiting all people and driving business success. Our new introduction presents the energy planning issue in simple and dynamic language.

2. Objective of the Study

This research focuses on how energy planning helps move away from fossil fuels to renewable power sources while meeting international Paris Agreement targets. Our analysis shows that good energy planning can help lower greenhouse gas emissions by 45% before 2030 and bring them to zero by 2050. Our research examines energy options and consumer trends to analyze how renewable energy affects society and the economy. Our work supports decision-makers by studying energy solutions to help them make policies that work for our planet and economy now and in the future (Suprayitno et al. 2024). This research stands apart because it examines all aspects of energy planning as the world moves toward renewable power sources.

The states of Tamil Nadu and Pune demonstrate the best ways to create sustainable energy systems in India. Tamil Nadu leads wind and solar power development with major installations such as the Kamuthi Solar Power Project. State policies drive renewable energy growth through investor-friendly guidelines, as Chandarahasan and Sheeba Percis (2023) explain. The state promotes energy efficiency while operating its advanced grid system to deliver continuous power supply. Pune works to reduce carbon emissions and stabilize energy delivery by placing solar panels on public infrastructure and updating energy distribution technology. Tamil Nadu and Pune prove that targeted energy planning produces economic growth, job opportunities, environmental protection, and enhanced social quality of life. Effective strategic planning leads us toward an energy system that works for tomorrow. Our study stands out by examining how energy planning affects technology, society, and the environment equally. This study examines how energy systems move from central distribution to localized sources while demonstrating why stakeholder collaboration matters. The study shows how Tamil Nadu and Pune successfully use projects to grow their economy while saving the environment. The document explains how to incorporate green hydrogen power and electric vehicles into today's energy frameworks to develop advanced sustainable energy methods.

3. Background Literature

The imperative for a global energy transition towards renewable sources is widely acknowledged due to pressing environmental concerns and the goals outlined in the Paris Agreement. The Intergovernmental Panel on Climate Change (IPCC) emphasizes the critical need for rapid reductions in greenhouse gas (GHG) emissions to prevent severe global warming, targeting a 45% decrease by 2030 and net-zero emissions by 2050 (Shahzad et al. 2024). This transition necessitates a departure from traditional fossil fuels—coal, oil, and gas—which currently account for over 80% of the world's energy supply and are major contributors to pollution.

Moving to clean energy, such as solar and wind power, offers significant environmental benefits, including improved air quality and reduced ecosystem damage, alongside economic and social advantages like job creation and enhanced public health (Algarni et al. 2023). However, achieving this transition is complex and requires robust energy planning. Effective energy planning involves understanding the intricate relationships between socioeconomic frameworks and energy systems. It provides a strategic framework for evaluating resources, simulating future energy scenarios, and developing sensible routes for energy transitions. Such planning must be comprehensive and inclusive, engaging governments, businesses, communities, and individuals.

Decentralized energy systems, while promoting new partnerships among stakeholders, also necessitate infrastructure enhancements. Research indicates that successful energy strategies integrate renewable energy systems, build reliable power grids, and foster collaboration among various partners (Koirala et al. 2018). The transition also carries social implications, including shifts in land use, changes in job markets as fossil fuel industries decline, and new opportunities in renewable energy sectors. Sustainable development underpins the global energy change, encompassing broader societal objectives such as poverty reduction, increased electricity access in underdeveloped areas, and social justice (Samarakoon 2019). The renewable energy sector itself is a significant source of employment, from manufacturing and installation to research and development.

In India, social, economic, and environmental benefits are driving the shift towards renewable power. India's heavy dependence on coal has historically led to high greenhouse gas emissions, warmer temperatures, and severe weather (Tiwari et al. 2023). To address these environmental issues, the Indian government is investing in eco-friendly energy sources to reduce carbon emissions and pollution. The declining costs of solar and wind energy have also made them financially attractive alternatives to fossil fuels.

The energy transition also necessitates technological innovation. Advancements in materials science, engineering, and information technology are required for the large-scale development and application of renewable energy solutions (Nižetić et al. 2023). The creation of novel energy storage systems, such as hydrogen fuel cells and batteries, is crucial for overcoming the intermittency of renewable energy and ensuring a continuous supply. Policies and regulatory structures are equally important for a successful energy shift, with governments enforcing laws that favor renewable energy and limit fossil fuel use. International cooperation, involving organizations like the International Renewable Energy Agency and the International Energy Agency, is also essential for sharing technical knowledge, linking power grids, and fostering scientific research.

4. Research Methodology

The research employed a mixed-methods approach, combining a comprehensive review of existing literature, case studies from two Indian states (Tamil Nadu and Pune), and analysis of news articles to determine current energy trends and the social and economic effects of renewable energy. This methodology allowed for a multifaceted understanding of energy planning and transition. The literature review formed the foundational component, drawing upon a wide range of academic papers, reports, and studies related to energy planning, renewable energy adoption, climate change, and sustainable development. This involved examining the importance of energy planning in achieving Paris Agreement goals, the necessity of reducing global greenhouse gas emissions, and the role of successful energy strategies in connecting renewable energy systems and building dependable power grids. The review also delved into the socioeconomic benefits of renewable energy development, such as employment generation and public health improvements. Key theoretical frameworks, including those related to integrated energy planning, policy results, technological development, and community participation, were explored. The research also considered the role of national governments in coordinating decarbonization efforts and the ambitious targets for global emissions reduction. The volatility of world energy markets and geopolitical concerns about fossil fuels were also investigated as drivers for energy security through diversification with renewables.

To provide practical context and validate theoretical findings, the study incorporated case studies from Tamil Nadu and Pune, India. These regions were selected as examples of best practices in creating sustainable energy systems. For Tamil Nadu, the research examined its leadership in wind and solar power development, including installations like the Kamuthi Solar Power Project, and the role of state policies in driving renewable energy growth. The study also considered the state's efforts in energy efficiency and advanced grid systems. For Pune, the methodology involved analyzing its initiatives to reduce carbon emissions and stabilize energy delivery, specifically focusing on solar panel installations on public infrastructure and updates to energy distribution technology. The case studies aimed to demonstrate how targeted energy planning in these regions has led to economic growth, job creation, environmental protection, and enhanced social quality of life. The overall approach aimed to provide a comprehensive understanding of the multifaceted aspects of energy planning and its role in successful energy transition, supporting policymakers in designing effective energy solutions. The methodology also aimed to analyze how renewable energy could support sustainable development objectives through its societal influence, and how different governmental sectors must collaborate to improve carbon emission reduction strategies.

5. Energy Transition and its Requirement

Our energy creation methods are transforming throughout the world. Our energy production is shifting from oil, gas, and coal usage toward renewable sources like sunlight, wind, and Earth-generated heat (Harlan and Baka 2024). These changes support climate protection while making the air cleaner and building energy solutions that endure. Our planet suffers damage when we fuel power plants using fossil fuels. Greenhouse gases raise Earth's temperature and bring harmful storms and floods that damage ecosystems. That is why we must switch to clean energy sources that do not cause these problems (Kumar et al. 2021, Yoro and Daramola 2020). Governments, international organizations, and local communities recognize how urgently carbon emissions should be lowered, and a switch to greener energy sources is needed (Tian et al. 2022). Apart from slowing down global warming, the change toward renewable energy sources seeks to reduce the adverse effects of traditional energy sources on air and water quality.

Toxin-producing combustion of fossil fuels threatens ecosystems and public health. Solar, wind, and hydropower, among other cleaner substitutes from renewable sources, help to minimize the environmental effects of energy generation and lower air pollution (Paravantis 2022). Solar and wind energy have shown remarkable cost savings, which makes them even more appealing substitutes for traditional fossil fuels. Apart from long-term sustainability, governments and businesses see the financial advantages of investing in renewable energy to generate employment and foster economic development (Wu et al. 2024). Energy security rests on the volatility of world energy markets and geopolitical concerns about fossil fuels. Countries depending on one or a few energy sources risk having their supplies disrupted and their pricing changed. Turning to renewable energy provides a means to diversify energy sources, hence lessening reliance on a few geopolitically sensitive resources. Energy security is promoted by constructing a more robust and dispersed energy infrastructure through diversification (Reeves 2012). Underlying the fundamental need guiding the worldwide energy change is the search for sustainable development. Apart from environmental and financial difficulties, renewable energy fits broader societal objectives, including decreasing poverty, increasing access to electricity in underdeveloped areas, and growing social justice. The project's primary goal is to examine renewable energy's contribution to reducing greenhouse gas emissions, thereby reducing social inequities and improving the quality of life for many populations through inclusive and sustainable development.

India leads the worldwide movement toward renewable power sources because of social requirements along with business and environmental benefits. To overcome climate change while making energy systems stronger and more sustainable, the nation puts more effort into solar and wind energy projects. India's power generation industry's heavy coal dependence created severe greenhouse gas emissions that resulted in warmer temperatures and severe weather, according to Bhatti et al.'s (2024) study. To address environmental problems, the Indian government and its partners invest in eco-friendly energy sources as part of their carbon emission reduction strategy. Renewable energy efforts in India help protect the environment and are focused on making air and water cleaner (Kushwah et al. 2024). The use of fossil fuels has become the major source of pollution that endangers both nature and human health. Through solar, wind, and hydropower systems, India works to decrease environmental damage from power generation and reduce the pollution in its air. The use of renewable energy now makes financial sense because of better technology and improved production methods. Solar power and wind energy have become low-cost solutions as their prices dropped in contrast to regular fossil-based fuels. The energy shift benefits our future resources and builds better economic conditions through more jobs and financial growth. India's renewable energy industry creates many jobs through its work in creating new products, setting up systems, and performing scientific studies. India's push for renewable energy solutions comes from its need to protect its energy supply (Chopdar et al. 2024). By using more renewable power sources, India gains control over its energy supply and reduces exposure to international supply problems. Using different types of renewable energy helps India build stronger and independent power systems. The main purpose of this energy shift is to achieve enduring progress. Renewable energy helps reach national objectives of fighting poverty and giving everyone fair access to electricity while empowering rural areas. To create a better future for its people, India increased renewable energy use while developing policies that help everyone prosper.

The study can fully meet the social need by using social and economic analysis of the acceptance of renewable energy. This implies community development, employment generation, and, more generally, social advantages of switching to renewable energy sources. Through an analysis of the whole influence on society, the study seeks to shed light on how renewable energy could support sustainable development objectives. Complementing the goal of fostering sustainable development, the paper investigates the inclusion and equity of renewable energy legislation. It looks at how these approaches view and control socioeconomic inequities such that the energy revolution could help many groups. Driven by needs anchored in environmental stewardship, economic viability, and social equality, the world energy landscape finds a tipping point. Turning to renewable energy sources is a multifarious endeavor meant to solve climate change, improve energy security, and advance sustainable development instead of only a technical one (Sovacool et al. 2021a). From manufacturing and installation to research and development, the renewable energy sector has grown to be a significant source of employment, with vocations ranging widely. Governments see the possibility for job creation and economic growth as they invest in clean energy infrastructure, thus promoting a transformation compatible with both financial and environmental goals. Since energy security is a long-standing issue for countries reliant on imported fossil fuels, it is of particular importance in the shift to renewable energy. By cutting reliance on foreign gas and oil, a country can increase its energy sovereignty and guard itself from geopolitical problems and supply interruptions. Using renewable and indigenous energy sources helps countries offer a steadier and more self-sufficient future. The several reasons and objectives of the energy transition address environmental, financial, and security issues as well as others. The necessity to solve climate change, the economic viability of renewable energy, and the pursuit of energy security define essential elements affecting the change toward greener energy sources. As countries embrace renewable energy, they are changing the world energy scenario and creating a more robust and sustainable future.

6. Importance of Energy Transition and System Transformation

It is difficult to overestimate the relevance of energy transition and system transformation in today's ever-changing environment. Globally, reliance on fossil fuels has resulted in serious environmental problems, including air pollution, climate change, and the destruction of natural ecosystems. Turning to renewable sources and adjusting energy systems will assist in solving problems and guarantee a sustainable future (Hoang and Nguyen 2021). First in importance is the energy reform as it aids in slowing down world temperature rise. Among other fuels, burning coal, oil, and natural gas accounts for most greenhouse gas emissions. By storing heat in the atmosphere, these pollutants cause global warming and disturbance of the surroundings. Turning to solar, wind, hydro, and geothermal energy sources will help to lower carbon emissions significantly. Renewable energy sources are better and more sustainable since they emit little to no greenhouse gases during operation (Evans et al. 2009). This shift will help us to limit the global temperature increase and stop the worst effects of climate change: loss of biodiversity, sea-level rise, and catastrophic storms. It is essential for enhancing public health, improving systems, and transforming energy. Burning fossil fuels generates hazardous pollutants, including nitrogen oxides, sulfur dioxide, particulate matter, and volatile organic compounds. These toxins, exacerbating lung and cardiovascular disorders, cause early mortality and lower quality of life. Using greener energy sources will help to greatly lower air pollution, thereby improving public health. For example, solar and wind energy enhance air quality and local conditions by generating electricity without generating hazardous emissions. Furthermore, switching to renewable energy helps to lessen reliance on limited resources and enhance energy security. Renewable energy sources are easily available, whereas geopolitical conflicts and price instability result from the limited and unequal distribution of fossil fuel sources worldwide. Changing the energy mix and supporting renewable technologies can help countries guarantee steady energy costs, lower their need for imported fossil fuels, and boost their energy independence. By establishing new businesses in the renewable energy sector, this change also promotes economic resilience and thereby helps stability and economic progress. Moreover, system reform is required to include renewable energy in the present infrastructure. Designed for centralized fossil fuel-based power generation, the conventional energy system has to be changed to welcome distributed and intermittent renewable energy sources (Soares et al. 2018). This change is part of upgrading the grid infrastructure, applying smart grid technology, and increasing energy storage capacity. Smart grids guarantee a constant and dependable energy supply through real-time monitoring and control of energy demand and supply, optimizing the use of renewable resources. Besides enhancing public health and reducing climate change, the energy transformation is essential for preserving natural ecosystems (Reddy et al. 2014). Energy storage technologies, including batteries, are vital for balancing the intermittency of renewable energy, storing additional energy during high generation, and releasing it. Damage to habitats, water contamination, and soil degradation are among the main effects of burning and using fossil fuels on the surroundings. For instance, whereas coal mining can seriously harm vast stretches of land, oil spills can have terrible consequences for marine life. Conversely, renewable energy sources have far less impact on their surroundings. While hydroelectric and geothermal electricity can be used in ways that protect natural ecosystems, installing solar panels and wind turbines will help to reduce disturbance of the nearby environment. Using renewable energy will help us to reduce the global environmental impact of our energy use and safeguard world biodiversity. Long connected to serious social and environmental injustices, including people's uprooting, worker exploitation, and local environmental deterioration, is the fossil fuel sector (Sovacool et al. 2021b). Employing more equitable and sustainable energy systems and switching to renewable energy will help to offset these inequalities. Projects using renewable energy can generate new economic prospects, including employment in installing and maintaining these systems, as well as supplying surrounding towns with consistent, reasonably priced, and clean electricity. Moreover, renewable energy projects can be conceived and carried out in a way that respects the rights and needs of surrounding populations, guaranteeing equal distribution of the advantages of the energy revolution. Additionally, necessitating great technological innovation are the developments in energy systems and the shift to renewable sources. Large-scale development and application of renewable energy solutions demand, among other areas, improvements in materials science, engineering, and information technology. For instance, developments in materials science will enhance solar panel and wind turbine affordability and efficiency; developments in information technology and grid management will help bring renewable energy into the grid. Moreover, the creation of novel energy storage systems, such as hydrogen fuel cells and batteries, will assist in overcoming the intermittency of renewable energy and offer a continuous supply. We can propel the technological innovation needed to support the energy transition and system change by employing research and development. Moreover, policy and regulatory structures are quite important factors for system transformation and energy transition. By enforcing laws favoring renewable energy and limiting the use of fossil fuels, governments and regulatory authorities can help to establish the environment for a successful energy shift (Sawin 2012).

India's use of renewable energy is changing the country's energy sector and solving many problems. A good example is the solar power program in Gujarat, where the Charanka Solar Park has become one of the biggest solar parks in Asia. This project generates much clean energy and reduces the area's dependence on coal and other non-renewable energy sources, which helps reduce harmful gases in the air. The solar park has made the air cleaner and improved people's health in the area by reducing dangerous pollutants that usually come from burning coal and oil. It has also created many jobs and boosted the local economy. This shows how renewable energy projects can benefit society, the economy, and the environment simultaneously. By investing in renewable energy projects like this, India is moving toward a more sustainable and robust energy future.

Using renewable energy sources and modernizing our energy infrastructure will help us provide a safe, healthy future for the next generations. Public understanding and involvement, supportive policies and legal frameworks, technical creativity, and international cooperation are all needed here. Apart from mitigating climate change and enhancing public health, the energy transition and system transformation are also crucial for maintaining natural ecosystems, supporting social and economic justice, and ensuring energy security. This is the moment to act; the advantages of system change and the energy revolution are unambiguous. Cooperation will guarantee a sustainable energy future and will allow us to improve the planet for all.

7. Energy Planning and its Ingredients

Modern society depends mainly on energy planning, which consists of several policies and methods meant to supply future energy demands while guaranteeing sustainability, economic advancement, and environmental preservation. To deliver consistent and reasonably priced energy services, it uses a whole strategy in handling infrastructure, technology, and energy resources. Effective energy planning has several parts with mixed technical, financial, environmental, social, and policy implications. Estimating energy usage is one of the essential components of energy planning. Accurate future energy demand projections will guarantee that energy supplies follow consumption patterns. Examining historical data, socioeconomic patterns, technological advancements, and population increase is part of this. Advanced modeling methods and simulation tools are used several times to project future energy usage under different scenarios. Understanding the factors influencing energy usage helps planners create plans to lower waste, increase energy efficiency, and manage demand. Thus, another essential element of energy planning is an analysis of the energy supply. This implies assessing several energy sources' environmental friendliness, dependability, and availability (Mourmouris and Portfolios 2013).

Many energy systems based on traditional fuels such as coal, oil, and natural gas have been developed, yet their limited character and environmental consequences call for a change to more environmentally friendly solutions. Because of their minimal environmental impact and chances for sustainable use, renewable energy sources, including solar, wind, hydro, and geothermal, are included in energy plans ever more. Besides their ability to satisfy current energy demand, planners have to evaluate the technical and economic feasibility of several sources. Including renewable energy among the fuels calls for a comprehensive evaluation of the infrastructure supporting them. Usually built for centralized fossil fuel-based power generation, the current energy system has to be changed to allow distributed and intermittent renewable energy sources. This suggests modernizing the system architecture, applying innovative grid technologies, and boosting capacity for energy storage. Smart grids guarantee a continuous and dependable energy supply through real-time monitoring and control of energy demand and supply, optimizing the use of renewable resources. Balancing the intermittency of renewable energy, holding extra energy during times of high generation, and releasing it when demand peaks depends greatly on energy storage technologies such as batteries. Another crucial instrument in energy planning is still economic analysis. Through proper analysis of energy phases from generation to delivery, we ensure consumers get energy services at reasonable rates. We need to measure how much different energy sources cost to run and maintain, including their fuel expenses, according to Santoyo-Castelazo and Azapagic (2014). Researchers analyze how different energy projects can create jobs and drive both economic growth and industrial progress. To achieve lasting sustainability, energy planners need to find a middle ground between keeping energy affordable and funding new technology and infrastructure investments.

Pune demonstrates smart energy strategies as an Indian city. The city created a complete strategy to deliver sustainable power solutions for future development while supporting both economic growth and environmental protection. The city of Pune now uses solar and wind technology as part of its energy sources to cut down on fossil fuel usage. The Pune Municipal Corporation has fitted public buildings across the city with numerous solar panels that create cleaner power and help reduce carbon emissions. To maintain consistent power supplies despite renewable energy variations, Pune has put money into advanced grid technology solutions. Research proves that these projects create employment opportunities and bring new money to the area which makes local businesses thrive. The city evaluates environmental impacts to direct efforts that reduce harm to nature and protect human wellness. Making the public part of the planning process and sharing benefits fairly between residents creates stronger public trust and leads to better health results. Through smart use of modern solutions and strategic planning, plus local participation, Pune is building a secure energy path for the future.

Countries need to work together and support energy planning projects to reach successful energy change all over the world. The International Renewable Energy Agency and the International Energy Agency lead global partnership efforts in energy planning. Energy planners need to use thorough methods to handle problems in infrastructure development, new technology integration, and management of energy resources. To achieve successful energy planning, teams need to study energy needs while exploring supply alternatives before they integrate infrastructure and evaluate economic and environmental effects along with social implications. They must build supportive policies that create energy resilience, encourage innovation, and work with international partners. Nations that work together can transfer useful information about energy systems to build better and stronger power infrastructures worldwide. Developing nations need these partnerships since they drive worldwide cooperation and boost faith in nations while delivering financial resources for clean energy change.

8. Role of Energy Planning in Transition and its Importance

Energy planning helps us build a future that protects nature while boosting our economy through sustainable energy transformation. The need for energy planning is extremely important because resource depletion and climate change threaten our entire planet (Stremke et al. 2011). Successful energy planning gives future generations dependable, low-cost power access while guiding our shift from fossil fuels into sustainable energy methods. The comprehensive planning process studies today's energy usage, then estimates how much we will need and designs environmentally friendly ways to supply it. The energy process needs policy creation alongside technical reviews and evaluations of available assets and investment strategies. Energy planning uses a complete strategy to direct choices while making the most of resources and protecting the environment. Renewable energy choices form the foundation of moving toward sustainable power networks. Each region has different levels of effectiveness for solar, wind, hydro, and geothermal energy, but all show strong potential for producing renewable power. Through detailed area studies, energy planning selects the best renewable power source available based on local environmental factors. Proper use of renewable resources ensures they provide the best possible input into the energy mix (Vidal-Amaro et al. 2015).

Effective energy planning and accepted renewable energy targets guide this transition. Government officials require complete energy strategies to set clear renewable energy targets along with specific implementation dates for current systems (Hoang et al. 2021). Those working on energy transformation projects use these defined targets as their roadmap and quality benchmarks for progress. The energy sector grows better when energy planning provides investment incentives through renewable energy policies such as feed-in tariffs, tax benefits, and financial aid. Decisions about energy support both energy source development and efficiency improvements across multiple sectors. Urban planners can develop strategies to make homes, businesses, factories, and transportation use less energy. Energy planners can promote energy conservation by investing in superior insulation materials, upgrading equipment, redesigning industrial operations, and expanding public transportation services. Our energy system becomes more secure and economical when these measures reduce pollution while saving energy. During the switch to alternative energy sources, energy planning maintains system stability and dependability. Solar and wind power systems create energy supply problems because they operate only during specific conditions. Careful energy management guarantees us a thriving economy while protecting our natural world. Considering our resource scarcity and climate concerns now requires urgent action. Our energy plan builds a stable, low-cost power system for future generations and helps us replace fossil energy sources. Through energy planning, we study current energy usage patterns to forecast future needs and develop environmentally friendly solutions. The planning process consists of forming regulations while analyzing technology capabilities alongside resource evaluation and investment scheme development. Effective energy planning supports wiser choices throughout resource allocation to lessen environmental damage. Better energy planning shows us ways to discover renewable power options while designing more sustainable power systems.

Tamil Nadu shows how smart energy plans can develop sustainable energy solutions across India. The state created an all-in-one energy plan to promote renewable power as a way to fight resource shortages and warming temperatures while supporting economic and environmental needs. The area of Muppandal in Tamil Nadu functions as an important wind power center, where its big wind farms produce substantial clean energy. Tamil Nadu put major resources into creating the Kamuthi Solar Power Project, which ranks as one of the biggest solar facilities globally. The government supports renewable energy investment using financial benefits and tax cuts. The state works to make its energy systems smarter in homes and workplaces to use less power and decrease pollution. Through advanced technology for electricity grids and energy storage, Tamil Nadu produces a dependable power supply despite changes in renewable energy production. The full strategy makes energy secure in the state while providing new jobs and making money locally to improve community quality of life. Tamil Nadu proves that smart planning plus targeted funds build a strong, lasting energy system.

The energy transition affects all nations, which requires them to establish cooperative partnerships. Countries join forces through energy planning to create agreements that let them share their technical knowledge and expertise. This helps everyone grow and develop. Many projects link power grids throughout different regions and build renewable energy partnerships between nations for scientific research. Energy planning guides the world toward adopting sustainable power solutions. Nations benefit from knowledge exchange to improve energy collaboration. Our battle against climate change requires a solid focus on energy planning strategy. Our energy policy efforts must confirm their support of both worldwide climate objectives and national environmental targets. In 2015, countries signed the Paris Agreement. This signed agreement establishes the main path for worldwide temperature reduction activities. The plan directs nations on energy strategies to combat global warming based on research by Höhne and colleagues in 2017. Energy affordability programs, including social tariffs and specific subsidies, contribute to the solution.

Energy planning must address how to manage growing energy needs. The growing need for energy caused by more people moving into cities and economic expansion creates uncertainty about our energy system's future. Energy planning is the development of demand-side management methods designed to maximize energy use and lower peak demand. It involves backing programs that let users manage their energy needs, combined with technologies that use less energy and ways to help people adjust their energy habits. Through strong energy planning models, we see technical progress and fresh innovations emerge. Our shift to renewable energy sources will drive us to create and adopt fresh technologies and ways of doing things. Energy planning identifies priority research areas and channels funds to drive an environment that promotes innovative thinking. This effort focuses on bringing public-private teams together while boosting innovation centers and backing new renewable energy business leaders. A good energy plan both converts energy systems to sustainability and carries out multiple operational tasks. It sets the plan for swapping fossil fuel energy infrastructure with renewable power systems.

Effortful management and research guide the strategic direction of energy planning, which meets the unique requirements of energy transformation. Through energy planning, we find the best ways to make renewable energy work. The method optimizes energy efficiency while guaranteeing power system stability and takes into account how energy solutions impact society and financial systems. Energy planning enables global cooperation while ensuring that environmental, social, and technological progress align with climate targets plus simpler access to energy for all communities. Energy planning brings together expert knowledge from various fields to show us the route toward fair and sustainable energy solutions for every person. The global energy demand continues to rise as we work to move toward renewable energy sources. Good, dynamic energy planning plays an essential role in shaping our future. Our detailed plans support us in managing energy system change while building a safer future for future generations.

9. Key Findings of the Study

a) ***Significant Reduction in Emissions*:** Our research shows that using renewable energy will reduce harmful greenhouse gas emissions in a major way. Our research found that renewable energy helps us achieve the targets set by the Paris Agreement of reducing emissions by 45% by 2030 and reaching net-zero by 2050.

b) ***Role of Energy Planning*:** Smart energy planning helps us build new renewable technologies and solve the complex distribution issues of distributed power systems. The system requires infrastructure improvements to manage its reliability and security with increased renewable energy sources.

c) ***Socioeconomic Benefits*:** The research shows renewable energy adoption produces economic benefits through job creation while improving public health and providing electricity to underserved regions.

d) ***Stakeholder Engagement*:** The results show that energy planning needs active engagement from public agencies, businesses, and local groups to create an accessible path for all participants.

10. Implications of the Study

a) ***Policy Development*:** The research provides significant findings to help decision-makers develop strong energy plans that protect the environment and solve wealth distribution problems. The study shows how different governmental sectors must work together to improve carbon emission reduction strategies.

b) ***Investment Mobilization*:** The research demonstrates renewable energy advantages while supporting investments in cleaner power systems and infrastructure to help us meet sustainability targets.

c) ***Framework for Future Research***: The results support future study about how energy system changes affect both society and the economy in developing areas with limited energy access.

d) ***Encouragement of Local Solutions***: When communities develop their own energy solutions, they bring unique ideas that make planning more effective and sustainable.

11. Discussion on Analysis

The analysis undertaken in this study highlights the critical importance of energy planning in achieving the ambitious goals of the Paris Agreement, particularly the reduction of global greenhouse gas emissions by 45% by 2030 and reaching net-zero by 2050. The findings strongly support the notion that a well-structured and comprehensive energy plan is not merely a technical exercise but a strategic framework that integrates technological advancements, policy initiatives, and community engagement to drive a successful transition to renewable energy.

One of the central themes emerging from the analysis is the significant potential of renewable energy to reduce harmful greenhouse gas emissions. By shifting away from fossil fuels, which are responsible for the majority of these emissions, to cleaner sources like solar, wind, hydro, and geothermal power, a substantial decrease in the carbon footprint can be achieved. This transition directly contributes to mitigating global warming and its adverse effects, such as extreme weather events and biodiversity loss.

The analysis underscores the multifaceted benefits of renewable energy adoption beyond environmental protection. Economically, the transition stimulates job creation in various sectors, from manufacturing and installation to research and development. This fosters economic growth and provides new opportunities, particularly in regions transitioning away from fossil fuel industries. Socially, the shift to renewables leads to improved public health by reducing air and water pollution associated with fossil fuel combustion. Furthermore, renewable energy initiatives can increase access to electricity in underserved areas, promoting social equity and contributing to broader sustainable development goals.

A key finding from the analysis is that effective energy planning is crucial for managing the complexities of integrating distributed power systems and ensuring grid stability and security as renewable energy sources increase. This requires significant infrastructure improvements, including the modernization of grid architecture, the application of smart grid technologies, and increased energy storage capacity. The intermittency of some renewable sources like solar and wind necessitates advanced management systems to guarantee a constant and reliable energy supply.

The case studies of Tamil Nadu and Pune vividly illustrate the practical application and success of strategic energy planning in India. Tamil Nadu's leadership in wind and solar power development, supported by investor-friendly policies and a focus on energy efficiency, demonstrates how comprehensive planning can drive significant renewable energy growth and provide a dependable power supply. Similarly, Pune's initiatives in public solar installations and grid technology updates highlight how local efforts can effectively reduce carbon emissions and stabilize energy delivery. Both examples show that targeted planning results in tangible economic growth, job creation, environmental protection, and an improved quality of life for communities. The analysis also emphasizes the importance of stakeholder engagement. Successful energy planning requires active collaboration among public agencies, businesses, and local communities to create inclusive pathways for all participants. This engagement ensures that energy solutions are tailored to local needs and foster public trust.

12. Conclusion

Recent studies explained how solar power combats climate change, but this research connects social details with system evaluations in its findings. This study demonstrates the beneficial relationship between technology and policy through quality energy planning instead of treating them as standalone research topics. Earlier research demonstrated renewable energy reduces emissions, yet this new research shows strategic planning helps maintain a stable power grid and engages community partners. Despite global research on renewable energy projects, this research targets Indian regional examples to show how local conditions affect energy transition solutions. The research provides new knowledge by showing how planning can simultaneously benefit the environment and society while also pushing for smarter energy system design to help us reach worldwide sustainable development objectives. The research demonstrates that energy planning helps move us away from fossil fuels to embrace renewable energy, which supports the Paris Agreement's climate objectives. When energy planning works well, it lowers our carbon footprint while creating jobs and improving people's health. The research shows in Tamil Nadu and Pune how companies must work with communities to find local solutions for energy development that grow economies and protect the environment. Organizations must build complete strategies to encourage renewable energy investments and create just energy system changes. Our next studies should look at how permanent outcomes develop from using renewable energy sources plus adding new technology in sustainability programs. By combining social impact analysis with technical data, this research presents a complete strategy for handling the challenges of changing energy systems.

References

Algarni, S., Tirth, V., Alqahtani, T., Alshehery, S. and Kshirsagar, P. 2023. Contribution of renewable energy sources to the environmental impacts and economic benefits for sustainable development. Sustainable Energy Technologies and Assessments 56: 103098.

Barton, H. 2009. Land use planning and health and well-being. Land Use Policy 26: S115–S123.

Bhatti, U. A., Bhatti, M. A., Tang, H., Syam, M. S., Awwad, E. M., Sharaf, M. et al. 2024. Global production patterns: Understanding the relationship between greenhouse gas emissions, agriculture greening and climate variability. Environmental Research 245: 118049.

Callister, P. and McLachlan, R. I. 2024. Managing Aotearoa New Zealand's greenhouse gas emissions from aviation. Journal of the Royal Society of New Zealand 54(4): 412–432.

Chandarahasan, C. and Sheeba Percis, E. 2023. The accessible large-scale renewable energy potential and its projected influence on Tamil Nadu's grid stability. Indonesian Journal of Electrical Engineering and Computer Science.

Chopdar, R. K., Sengar, N., Giri, N. C. and Halliday, D. 2024. Comprehensive review on agrivoltaics with technical, environmental and societal insights. Renewable and Sustainable Energy Reviews 197: 114416.

Evans, A., Strezov, V. and Evans, T. J. 2009. Assessment of sustainability indicators for renewable energy technologies. Renewable and Sustainable Energy Reviews 13(5): 1082–1088.

Fazey, I., Schäpke, N., Caniglia, G., Patterson, J., Hultman, J., Van Mierlo, B. et al. 2018. Ten essentials for action-oriented and second-order energy transitions, transformations and climate change research. Energy Research & Social Science 40: 54–70.

Friedl, C. and Reichl, J. 2016. Realizing energy infrastructure projects—A qualitative empirical analysis of local practices to address social acceptance. Energy Policy 89: 184–193.

Harlan, T. and Baka, J. 2024. Stacked energyscapes: Conceptualizing fossil fuel and renewable energy entanglements in low-carbon transitions. Energy Research & Social Science 115: 103648.

Hoang, A. T. and Nguyen, X. P. 2021. Integrating renewable sources into energy system for smart city as a sagacious strategy towards clean and sustainable process. Journal of Cleaner Production 305: 127161.

Hoang, A. T., Nižetić, S., Olcer, A. I., Ong, H. C., Chen, W. H., Chong, C. T. et al. 2021. Impacts of COVID-19 pandemic on the global energy system and the shift progress to renewable energy: Opportunities, challenges, and policy implications. Energy Policy 154: 112322.

Höhne, N., Kuramochi, T., Warnecke, C., Röser, F., Fekete, H., Hagemann, M. et al. 2017. The Paris Agreement: resolving the inconsistency between global goals and national contributions. Climate Policy 17(1): 16–32.

Koirala, B. P., van Oost, E. and van der Windt, H. 2018. Community energy storage: A responsible innovation towards a sustainable energy system? Applied Energy 231: 570–585.

Kumar, A., Nagar, S. and Anand, S. 2021. Climate change and existential threats. pp. 1–31. In Global Climate Change. Elsevier.

Kushwah, S., Iyer, R., Agrawal, A. and Korpal, S. 2024. Understanding switching intentions towards renewable energy technologies using push-pull-mooring framework. Journal of Cleaner Production 465: 142656.

Lew, D., Bakke, J., Bloom, A., Brown, P., Caspary, J., Clack, C. T. et al. 2021. Transmission planning for 100% clean electricity: enabling clean, affordable, and reliable electricity. IEEE Power and Energy Magazine 19: 56–66.

Maroto-Valer, M. M., Song, C. and Soong, Y. (eds.). 2012. Environmental Challenges and Greenhouse Gas Control for Fossil Fuel Utilization in the 21st Century. Springer Science & Business Media.

Mourmouris, J. C. and Potolias, C. 2013. A multi-criteria methodology for energy planning and developing renewable energy sources at a regional level: A case study Thassos, Greece. Energy Policy 52: 522–530.

Mundaca, L., Busch, H. and Schwer, S. 2018. 'Successful' low-carbon energy transitions at the community level? An energy justice perspective. Applied Energy 218: 292–303.

Nižetić, S., Arıcı, M. and Hoang, A. T. 2023. Smart and sustainable technologies in energy transition. Journal of Cleaner Production 389: 135944.

Paravantis, J. A. 2022. Socioeconomic aspects of third-generation biofuels. pp. 869–917. In 3rd Generation Biofuels. Woodhead Publishing.

Reddy, K. S., Kumar, M., Mallick, T. K., Sharon, H. and Lokeswaran, S. 2014. A review of Integration, Control, Communication and Metering (ICCM) of renewable energy based smart grid. Renewable and Sustainable Energy Reviews 38: 180–192.

Reeves, M. 2012. Movement, Power and Place in Central Asia and Beyond. London and New York: Routledge.

Rhodes, C. J. 2019. Only 12 years left to readjust for the 1.5-degree climate change option–Says International Panel on Climate Change report: Current commentary. Science Progress 102(1): 73–87.

Samarakoon, S. 2019. A justice and wellbeing centered framework for analysing energy poverty in the Global South. Ecological Economics 165: 106385.

Santoyo-Castelazo, E. and Azapagic, A. 2014. Sustainability assessment of energy systems: integrating environmental, economic and social aspects. Journal of Cleaner Production 80: 119–138.

Sawin, J. L. 2012. National policy instruments: policy lessons for the advancement and diffusion of renewable energy technologies around the World. pp. 71–114. In Renewable Energy. Routledge.

Scott, D. and Gössling, S. 2021. Destination net-zero: what does the international energy agency roadmap mean for tourism? Journal of Sustainable Tourism 30(1): 14–31.
Shahzad, S., Faheem, M., Muqeet, H. A. and Waseem, M. 2024. Charting the UK's path to net zero emissions by 2050: Challenges, strategies, and future directions. IET Smart Grid.
Soares, N., Martins, A. G., Carvalho, A. L., Caldeira, C., Du, C., Castanheira, É. et al. 2018. The challenging paradigm of interrelated energy systems towards a more sustainable future. Renewable and Sustainable Energy Reviews 95: 171–193.
Sovacool, B. K., Hess, D. J. and Cantoni, R. 2021a. Energy transitions from the cradle to the grave: a meta-theoretical framework integrating responsible innovation, social practices, and energy justice. Energy Research & Social Science 75: 102027.
Sovacool, B. K., Turnheim, B., Hook, A., Brock, A. and Martiskainen, M. 2021b. Dispossessed by decarbonization: Reducing vulnerability, injustice, and inequality in the lived experience of low-carbon pathways. World Development 137: 105116.
Stremke, S., Neven, C. and Boekel, A. 2011. Beyond uncertainties: how to envision long-term transformation of regions? International Conference on Industrial Technology.
Suprayitno, D., Iskandar, S., Dahurandi, K., Hendarto, T. and Rumambi, F. J. 2024. Public policy in the era of climate change: adapting strategies for sustainable futures. Migration Letters 21(S6): 945–958.
Tian, J., Yu, L., Xue, R., Zhuang, S. and Shan, Y. 2022. Global low-carbon energy transition in the post-COVID-19 era. Applied Energy 307: 118205.
Tiwari, V., Garg, A., Kapshe, M., Deshpande, A. and Vishwanathan, S. 2023. Assessing possibilities for coal continuance in India under climate constraints. International Journal of Greenhouse Gas Control 122: 103811.
Vidal-Amaro, J. J., Østergaard, P. A. and Sheinbaum-Pardo, C. 2015. Optimal energy mix for transitioning from fossil fuels to renewable energy sources—The case of the Mexican electricity system. Applied Energy 150: 80–96.
Wu, H., Feng, Z., Sun, T., Li, R. and Zhao, H. 2024. Efficiency, sustainability, and resilience a trifecta for a green economic recovery through natural resource markets. Resources Policy 88: 104435.
Yoro, K. O. and Daramola, M. O. 2020. CO_2 emission sources, greenhouse gases, and the global warming effect. pp. 3–28. In Advances in Carbon Capture. Woodhead Publishing.

Chapter 3

Integrative Approaches to Renewable Energy Transition in India

Vivek Sen[1] and *Romi Sainy*[2,*]

1. Introduction

The energy transition in India needs analysis of our society's needs alongside our technological capabilities and financial realities. The analysis identifies possible directions for transitioning the power generation sector to 100% renewable energy with green technologies by 2050. Our study investigates India's energy evolution, especially the relationships between its market trends, environmental conditions, and technological progress. This research stands alone in examining both opportunities and challenges across multiple sectors comprehensively.

India has ambitious national targets for energy generation from renewable sources, emerging from the increasing awareness of climate change. An energy system with minimal carbon emissions would alleviate the impacts of climate change and reduce adverse health effects associated with fossil fuel consumption. In contrast, renewable energy technologies are progressively viewed as a method to provide economical energy in the long run. India has significant potential for widespread adoption of renewable energy technologies due to its geographical characteristics. The difficult task of energy transition in India necessitates understanding social, technological, and economic issues. India has set high national targets for renewable energy output, motivated by concerns about climate change and energy security. A low-carbon energy system can potentially halt catastrophic climate change and mitigate the detrimental effects of fossil fuel use on human health. India's geographical location

[1] Climate Policy Initiative, India.
[2] Professor – Marketing, Jaipuria Institute of Management, Indore, India.
Emails: vivek.sen@cpiglobal.org
* Corresponding author: romi.sainy@jaipuria.ac.in

DOI: 10.1201/9781003475385-3

offers numerous opportunities to develop large-scale renewable energy sources. Low total renewables share and energy poverty in rural areas are supply and demand concerns that the shift must address.

Therefore, there is considerable expectation that renewable energy sources will emerge from their current peripheral state and gain a bigger role in Indian energy production. The low-carbon transition is mired by numerous barriers—both on the supply and demand sides. In the context of the supply side, the overall share of renewable energy technologies is rather low in India compared to its competitors. Renewables face various difficulties in being recognized as important players in the Indian energy system.

1.1 Current Energy Landscape in India

As one of the fastest-growing economies, India's energy needs are significant and continually expanding. Understanding this landscape requires a multifaceted examination of the various elements contributing to the nation's energy profile.

A mix of conventional and renewable sources largely meets India's energy consumption. Fossil fuels, mostly coal, constitute a significant segment of the energy portfolio due to the nation's extensive coal deposits. Recent years have witnessed a significant evolution toward renewable energy foundations, including solar, wind, and biomass, aligning with the global trend toward sustainability. The total installed capacity as of December 2023 was 441 GW.[1]

A diverse mix of sources characterizes India's installed capacity landscape. As of recent data, coal remains the dominant energy source, accounting for approximately 49% of the country's total installed capacity (Figure 1).[2] The reliance on coal poses

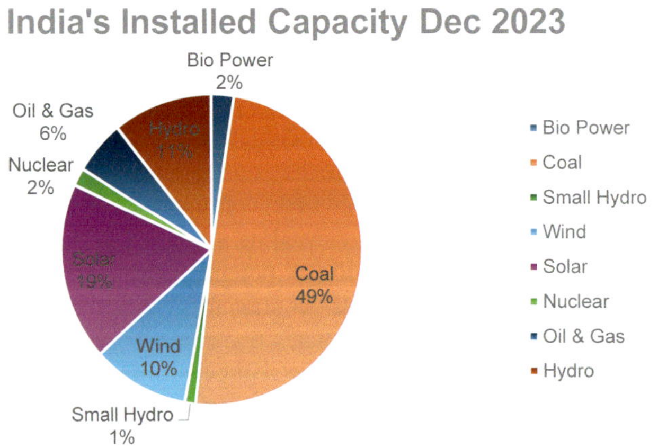

Figure 1. India's Installed Capacity. Source: Climate Policy Initiative.

[1] Government of India. (2024). RENEWABLE ENERGY RENEWABLE ENERGY STATISTICS 2023-24. Retrieved from: https://cdnbbsr.s3waas.gov.in/s3716e1b8c6cd17b771da77391355749f3/uploads/2024/10/20241029512325464.pdf.

[2] PIB2009196.pdf.

significant environmental concerns, particularly regarding greenhouse gas emissions and air pollution.

Natural gas, hydroelectric power, nuclear energy, and renewables, including solar and wind, constitute the other segments of India's energy mix. Oil and gas contribute about 6–7% of total energy consumption, but its potential remains underutilized due to infrastructural limitations.

Hydropower constitutes around 10–12% of the energy mix and is a crucial renewable source, benefiting from India's vast river systems. However, seasonal variations and ecological impacts limit its scalability. Nuclear, though a small part of the mix at about 2%, offers a reliable and low-carbon alternative, albeit one fraught with challenges regarding safety, waste management, and high capital costs. India, among the fastest-growing nations in the world, has increasing energy needs. Though mostly coal, the energy balance is dominated by fossil fuels; in recent years, there has been a notable change toward renewables. By December 2023, there was 441 GW of installed capacity overall. With 49% of capacity, coal still rules, while renewables are growing—solar at 19% and wind at 10% (Ang et al. 2022). Ten to twelve percent of the total is hydropower, with nuclear accounting for about two percent. Although government subsidies have helped renewables, especially solar and wind, greatly expand, the reliance on coal creates environmental concerns.

Renewable energy sources, particularly solar and wind, have shown remarkable growth. Solar energy, supported by government initiatives, has seen significant adoption, contributing around 19% of the total capacity. Wind energy, with its substantial potential, currently provides about 10%. Biomass and other alternatives, while not as prominent, play a supplementary role, especially in rural and semi-urban areas.

1.2 Current Policies and Regulations

Over the past ten years, the Indian government has developed its energy policies to create both an efficient and sustainable energy system for the nation. In 2008, the National Action Plan on Climate Change (NAPCC) started operating through eight missions, one of which is the National Solar Mission and the National Mission for Enhanced Energy Efficiency. The missions target climate change reduction through renewable energy promotion and better energy-saving methods.

Under the Electricity Act of 2003, the government introduced important rules to increase competition and welcome private investors into the energy marketplace. State electricity regulatory commissions emerged under the law and today help enforce local energy policies and set electricity prices. Since 2015, the Ujjwal DISCOM Assurance Yojana (UDAY) has worked to strengthen the finances of electrical distribution firms to enable more effective power service delivery.

Under Renewable Purchase Obligations, state electricity regulators must buy a set amount of energy from renewable sources. The GST Council decided to put renewable energy equipment in low tax ranges, which helps more people and companies invest in clean energy.

These laws and regulations jointly aim to provide a balanced and progressive energy ecosystem that fosters sustainable growth while directly confronting economic and environmental issues.

1.3 Future Trends in India's Energy Mix

India, a rapidly expanding economy, confronts the difficulty of reconciling its escalating energy requirements with the necessity for sustainable growth. India, being the world's second-most populous nation, is anticipated to experience an increase in energy demand in the forthcoming years, propelled by population growth and economic development (Paul et al. 2013). The nation's dependence on fossil fuels, especially coal, has resulted in considerable environmental issues, prompting an increasing acknowledgment of the necessity to diversify the energy portfolio and adopt renewable sources.

The nation's plentiful solar and wind resources, coupled with ambitious governmental objectives, are driving this shift. While fossil fuels, primarily coal, dominate the energy mix, the share of renewables is steadily increasing. This transition is propelled by a convergence of causes, including decreasing costs of renewable energy technology, increasing environmental concerns, and the government's advocacy for energy autonomy.

The shift to cleaner sources requires upgrading the Indian power network and installing battery storage and smart grid technology. The government is working to build a resilient hydrogen economy that functions as an ecological fuel carrier and energy reserve. Renewable energy sources create hydrogen that serves as both a transport fuel and an energy production resource (Figure 2).

India plans to greatly expand renewable energy through more solar and wind installations. The nation's renewable energy goals, combined with policy and new technologies, will speed up this energy transformation. By combining different energy sources with large amounts of renewable energy, India can protect its energy supply while cutting emissions and fostering economic growth.

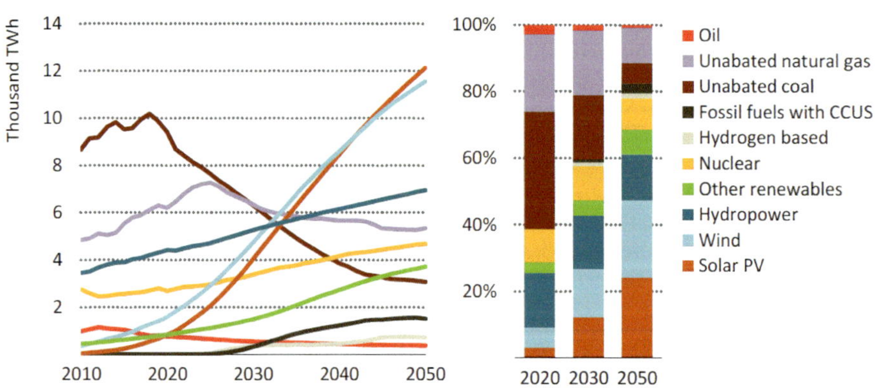

Figure 2. Energy Usage Projection. Source: International Energy Agency.

2. Literature Review

India's energy transition is taking place in a complex and dynamic environment. With an existing capacity of 174 GW anticipated for mid-2024 and roughly 200 GW reached by early 2025, the country is aggressively increasing its renewable energy capacity in order to meet the goal of 500 GW of non-fossil energy by 2030 (Reuters 2025, IIC 2024). On the other hand, significant barriers include outdated infrastructure, an overreliance on coal (above 70% of generation), and limited energy storage (AP News 2023, IIC 2024). Several studies and policy analyses concur that adding variable renewables, such as wind and solar, to India's coal-based grid will require large investments in grid flexibility, including storage, sophisticated control systems, and flexible regulations (IFRI 2025, IEEFA 2021, IEA 2021, 2020).

High renewable scenarios can be financially feasible, according to academic modeling, but they call for complex balancing techniques. For real-time balancing, Rahul Tongia's parametric model shows that relying too heavily on renewable energy sources without storage is insufficient; storage and demand response are necessary but currently expensive (Tongia 2022). As demonstrated by Mohan et al. (2021), hybrid solar-battery systems may offer baseload-like reliability while easing the phase-out of coal plants if they are paired with reasonably priced capital. Ding et al. (2024) show that retrofitting coal plants with carbon capture technology or biomass co-firing could help address the unequal distribution of renewable energy across states and enable India to decarbonize at a reasonable cost.

Technical issues include frequency stability and grid inertia. The rapid growth of inverter-based renewable energy sources reduces system inertia and complicates fault response, according to Debanjan and Karuna (2022), which calls for intelligent frequency management and artificial inertia solutions. Grid codes are changing to accommodate synchronous condensers and grid-forming inverters, but their actual application varies by state, according to The Electricity Sector in India (2025). Another IFPRI memo warns that grid stability within India's limited frequency range (49.90–50.05 Hz) may be threatened by inadequate storage capacity (4.7 GW pumped hydro, roughly 219 MWh BESS, compared to the required 60 GW by 2030) (IFPRI 2025).

Finance and policy are equally vital. Socio-technical frameworks that support intelligent grids, flexible coal operations, and equitable labor transitions are promoted by the draft Electricity (Amendment) Bill 2020 and scholarly analyses (Debnath et al. 2022). It is essential to promote demand-side flexibility and expedite the deployment of smart meters (GroundReport 2024, IEEFA 2021). Concurrently, large-scale projects like Gujarat's 30 GW hybrid park with 14 GWh storage show India's ambitions by combining advanced transmission systems, solar, wind, and grid expansion.

Notably, innovation is being propelled throughout the Global South by domestic models and collaborations. Battery energy storage system (BESS) pilot projects in India, like BSES Rajdhani–GEAPP, show how effective public-private-philanthropic partnerships are at delivering clean, dependable electricity to low-income communities while reducing intermittency through storage (Reuters 2025). Time-of-day pricing, demand response, and flexible coal dispatch are proven to improve grid performance across states, according to international models such as the IEA's India Integration Report (IEA 2021).

3. Drivers of Energy Transition

Different connected reasons drive India toward sustainable energy development. This section studies the important forces pushing India toward energy transition. Our first topic of study includes financial aspects such as savings from renewable energy options and how worldwide economic trends push the need for sustainable power. Our exploration of environmental drivers comes next because climate change now presents the greatest danger to our survival. New technology makes it possible to transform our goals into reality by helping us create renewable energy that works better and costs less.

These elements combine to make a strong model that explains how India shifts to new energy sources. Each part works alone but also strengthens the whole switch to clean energy. The decreasing costs of renewable energy sources, along with increasing fossil fuel prices, make it easier to create and enforce tougher environmental rules because countries want to meet their climate goals. At the same time, technological progress lowers expenses while making renewable energy easier to expand. Different forces work together to drive India's energy changes. Economic reasons include foreign fossil fuel price increases, lower renewable energy prices, and the fact that renewable energy creates jobs and receives state financial incentives. Major environmental goals deal with lowering planetary warming trends while fighting air pollution and protecting different living organisms. The growth of renewable energy thrives on improvements in solar PV systems and smart power networks.

Decision-makers and investors require this knowledge to develop successful energy transition plans for India. Our detailed analysis of the drivers below shows how their connections can help create both durable and adaptable energy systems.

3.1 Economic Factors

Multiple linked financial elements shape India's movement to new energy systems. India's need to switch to different energy sources results from how much the country spends on importing fossil fuels. As the world's third-largest oil importer, India allocates a major part of its GDP to buying energy from other nations. The rising import bills of fossil fuels show us why we must urgently move toward local renewable power generation that will provide stable prices.

The rapidly decreasing prices of solar and wind power technologies now make them affordable alternatives to traditional energy methods. According to IRENA's report, utility-scale solar power costs fell by 82% between 2010 and 2019. Clean energy projects become financially attractive because renewable energy operations now cost less. Both government agencies and private businesses respond by increasing their investment funds.

Energy transition will make multiple job types available. This economic plan matches India's targets to increase jobs while growing its economy. National financial programs and support determine key economic choices in society. Different incentive programs, including feed-in tariffs, tax benefits, and initial funding assistance, help investors move past the start-up costs of renewable energy projects. Financial benefits help India move quickly toward sustainable power solutions that also save money.

3.2 Environmental Considerations

Low economic development and environmental sensitivities require India to put a strong focus on environmental factors during its energy transition. More than 70% of India depends on fossil fuels for power generation, which creates major air pollution and greenhouse gas emission problems. The use of fossil fuels has harmful effects on our environment, so we must change to using sustainable energy options. Reversing climate change stands as the top reason for switching our energy systems. India has pledged to make its energy cleaner by lowering carbon output and growing renewable energy sources. The energy shift helps control global warming and builds safety measures to support at-risk communities.

Switching to solar and wind power helps solve the serious air quality problems hurting people in cities throughout India. Scientists show air pollution causes millions of early deaths every year in India. Using cleaner energy sources instead of coal and petroleum helps India protect its citizens' health and improve how people live. Protecting biological diversity forms a vital part of our environmental initiatives. Fossil fuel operations and large mining activities damage natural habitats and put animal lives at risk. Although renewable energy systems have some environmental costs, they remain the better choice because they produce smaller ecological footprints.

3.3 Technological Advancements

Technological advancements form an essential part of India's strategy for energy change. Our progress in renewable energy systems, smart grid technology, and enhanced energy storage shows major results. Modern technology makes renewable power systems both more productive and affordable. Solar photovoltaics (PV) technology shows important progress. Studies in materials science have helped solar PV cells perform better, especially through perovskite implementation. New technology now allows harvesters to extract more power from areas where sunlight is limited. The last ten years have brought major price reductions in solar panels, which put solar energy on par with other options. The energy storage sector has made major developments through better lithium-ion batteries and new technologies, including solid-state and flow batteries. The storage systems we need to stabilize power supplies from solar and wind energy sources help provide continuous electricity.

Smart grid technology deployment makes our power distribution network run better and stay more consistent. Networks that use intelligent monitoring enable us to adjust energy usage based on demand and watch power systems in real time while bringing together renewable energy of all sizes. Research shows how blockchain technology upgrades can simplify energy transactions and make them more transparent. New technology makes it easier for India to expand renewable energy systems while keeping the energy systems sustainable as they solve key integration challenges faced by the present electrical infrastructure. Technical innovations help India build better energy systems that work more efficiently while becoming sustainable and resilient. New developments in renewable power systems and energy management help India fulfill its power needs while lowering environmental damage.

4. Research Methodology

The chapter uses a descriptive methodology by assessing secondary data sources. The primary objective of the chapter is to explore the current status of energy transition in India, the recent trends, challenges, and opportunities faced by the energy sector. A descriptive method is suitable for assessing and summarizing the policy-level developments and initiatives taken by industry players, as well as studying various elements like financial, economic, technological, and environmental factors that impact the growth and reshaping of the sector to develop a coherent understanding of the present and future of energy transition in India.

Primary data sources used in the study were government reports (Ministry of Power, MNRE), international agency data (World Bank, IEA, and UNFCCC, Climate Policy Initiative), peer-reviewed articles published in reputed journals and magazines, and industry reports published during 2000–25.

5. Components of Energy Transition

Energy transition means updating our energy systems to fight climate change while supporting continuous economic development. Energy transition takes on special significance for India as it evolves through fast urban expansion and industrial development to become a nation with reduced carbon emissions (Figure 3). This section explores the key components of energy transitions across four major sectors: electricity, buildings, transport, and industry.

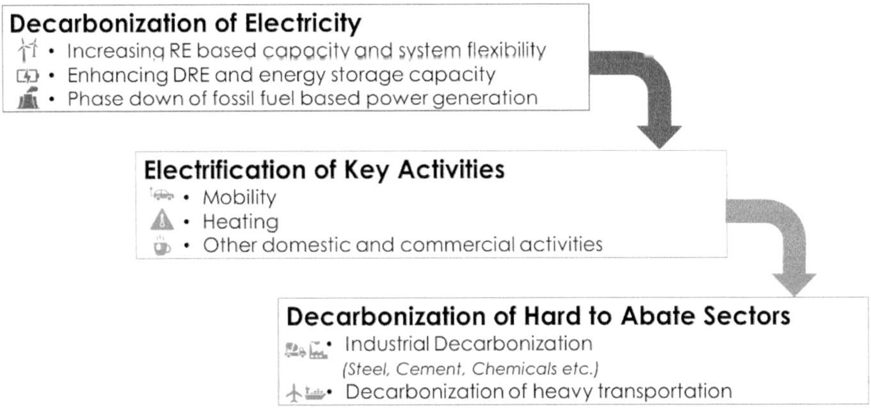

Figure 3. Components of Energy Transitions. Source: Climate Policy Initiative.

5.1 Electricity Sector

The electricity sector is pivotal in India's energy transition. Key components include:

- **Renewable Energy Integration:** India has set ambitious targets for renewable energy, aiming for 500 GW of renewable capacity by 2030.[3] Solar and wind energy are at the forefront of this transition.
- **Smart Grids:** The adoption of smart grid technology enhances the efficiency of electricity distribution, enabling better management of energy supply and demand.
- **Energy Storage Systems:** Energy storage solutions, like battery technology, are crucial for stabilizing the grid and tackling intermittency in renewable sources.

5.2 Building Sector

The building sector is a significant contributor to energy consumption. Transition components include:

- **Energy-Efficient Design:** Using green building designs with energy-saving features brings a substantial reduction in building power use.
- **Green Certifications:** The LEED and GRIHA standards promote efficient building work by certifying practices that lead to reduced energy consumption.
- **Retrofitting Existing Buildings:** When buildings receive updated energy-saving technology, it needs less power to run.

5.3 Transport Sector

The transportation system plays a vital role in decreasing greenhouse gas production. Key components include:

- **Electrification of Transport:** Electric vehicles must replace traditional cars as our top transportation solution. The government supports this movement through incentives and the construction of more charging stations.
- **Public Transportation Enhancement:** When public transit works well, people choose to use it instead of their cars, which helps save energy.
- **Promoting Non-Motorized Transport:** Cities can lower their energy use and emissions when they build separate paths for walkers and cyclists.

5.4 Industrial Sector

The industrial sector is a major energy consumer. Transition components include:

- **Energy Efficiency Improvements:** Introducing modern energy control systems along with new, efficient machines will decrease energy requirements for industries.

---

[3] 500 GW Nonfossil Fuel Target | Government of India | Ministry of Power.

6. Challenges to Energy Transition

The move toward energy transition in India demands detailed planning and study because many problems stand in its path. As India works to replace traditional energy with renewables, it faces several barriers that could slow this transformation. India faces interconnected difficulties related to financial issues, physical systems, government rules, and national strategies.

Stakeholders need to identify these problems so they can develop better action plans. Renewable energy projects create significant financial barriers because they require higher upfront costs than conventional energy solutions. Our current power grid and storage system cannot support renewable energy sources because they need upgrades to handle power level changes from solar and wind energy. India must deal with numerous problems as it changes its energy consumption practices. The high start-up costs, combined with risky industry perceptions, unfair subsidies, and the billions needed for investment, represent the main financial constraints. Three main infrastructure problems block progress: inadequate EV support systems, difficulties in acquiring land, and modernizing electrical grids. India's energy sector is limited by government policies, including only one central energy policy, plus administrative delays and underfunding of grid networks, which prevents renewable energy storage from growing.

Government rules and standards form a significant obstacle within the sector. India's government procedures and conflicting regulations in energy policy discourage private investment and delay renewable energy development. Outdated government policies do not reflect recent renewable technology improvements, creating problems for efficient energy system transformation. Renewable energy development depends heavily on both land ownership rules and public participation outcomes.

6.1 Financial Constraints

Many financial barriers affect India's energy transition and slow down renewable energy growth. Starting renewable energy facilities requires a lot of upfront investment. Getting started with solar, wind, or biomass energy projects requires large upfront costs, which limit the ability of many participants to enter the Indian market. The main problem is obtaining long-term loans at suitable interest rates. Traditional lenders see renewable energy projects as risky because they depend on emerging technologies and unpredictable governmental regulations. The negative industry opinion about renewable energy investments results in expensive financing options that discourage funds from flowing into clean energy projects.

The constantly changing nature of support programs adds more difficulty to financing decisions. Government programs supporting renewable energy projects suffer from inconsistent budget policy and slow payment processes, which disrupt developer and investor money flow. To move toward a low-carbon economy, India must make substantial investments. Research shows India needs trillions of US dollars to rebuild its energy system and embrace sustainable energy solutions.

According to the Council on Energy, Environment, and Water (CEEW), India will need USD 10.1 trillion to reach net-zero emissions by 2070.

By 2050, the Reserve Bank of India calculates that India needs USD 7.2 to 12.1 trillion to move forward with the energy transition. These enormous figures show how big a problem India must solve. The funds will support renewable power generation, grid systems, and energy storage development while advancing electric vehicle use and sustainable hydrogen production. Making this large investment requires both public and private organizations to work together toward a common goal. Strong government action and helpful financial support will help India attract funding from investors worldwide to support its clean energy needs. To solve this funding problem, we need to create more affordable investment options while defining automatic processes for subsidies and promoting creative ways to reduce investment risks in renewable energy projects.

6.1.1 Current State of Finance Availability

While the exact figure fluctuates based on various factors and methodologies, it is evident that the country needs trillions of US dollars in investment. The Climate Policy Initiative (CPI) estimates that India's climate action would require approximately USD 170 billion annually between 2015 and 2030, totaling USD 2.5 trillion (Figure 4). There are significant gaps in financing the energy transition in India and globally.

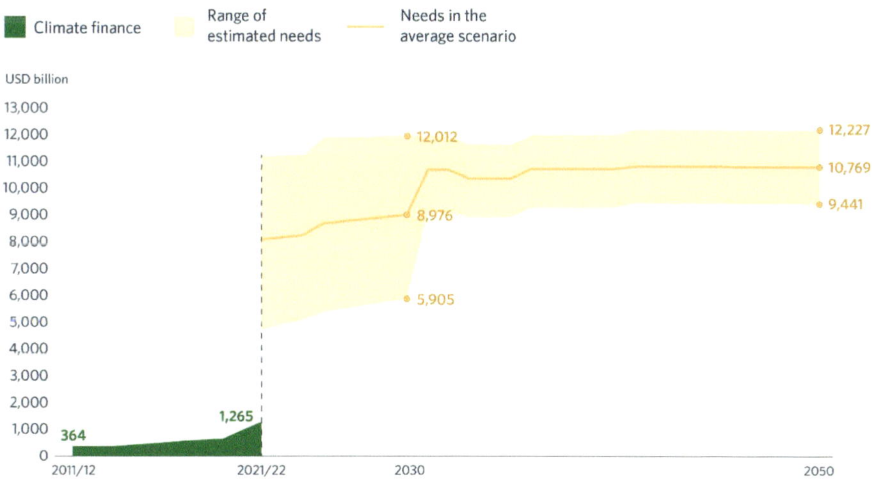

Figure 4. Energy transition financing in India. Source: Climate Policy Initiative.

6.2 Cost of Capital Constraints

Transitioning to a low-carbon economy necessitates substantial investments in renewable energy, energy storage, grid infrastructure, and other clean technologies. However, the cost of capital—the rate of return expected by investors—poses a significant hurdle to financing these projects. Each risk factor—sovereign, political,

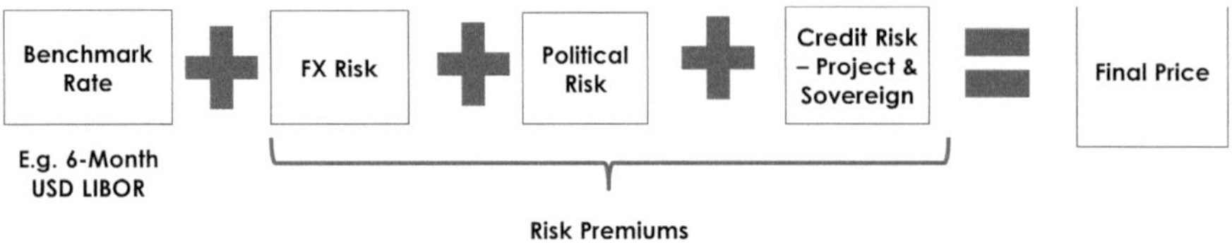

Figure 5. Cost of Energy Transition. Source: Climate Policy Initiative.

and business/project—has a certain price associated with it in the form of basis points, which get added to the cost of finance (Figure 5). Based on conventional finance theory, namely the Capital Asset Pricing Model (CAPM) and/or Risk-Adjusted Return on Capital (RAROC).

A number of reasons make capital more expensive for companies. Most energy transition projects test new technologies that come with specific risks. People looking to invest in these projects need better payback due to the risk of failure. Investors who want to make quick profits lose interest in projects that need to run for many years. Clean energy projects face heightened investment risks because regulations in this field shift frequently and remain complicated. An unstable financial background and limited financial resources in developing regions make funding more expensive. Higher costs of capital create multiple problems across different areas. Expensive capital either prevents projects from starting, reduces investment levels, or slows technological innovation. High costs for clean energy debt create disadvantages that help traditional energy projects compete against greener options. The move toward using low-carbon energy sources takes longer to achieve and comes at higher costs.

6.3 Infrastructural Constraints

Building infrastructure plays an essential role in India's move from traditional to modern energy systems. Today's energy system depends too much on coal power for India's energy needs to switch safely to different energy sources. Moving toward renewable power means India needs to enhance various parts of its power system significantly. To connect renewable energy sources with the national system, major improvements to the grid are required. The national power system struggles to handle the uneven production levels of wind and solar power sources. To stabilize our energy supply, we must improve existing grid technology and add smarter systems with better storage options.

Building renewable energy power plants requires smart planning and significant investment. Renewable energy infrastructure needs simpler procedures and strong government support to complete land transactions and transmission line setups. Outdated energy systems block the flow of electricity and create transition difficulties. Renewable energy systems for small locations enable remote areas to generate their own power supply and reduce demand on the main power grid. The growth of India's energy system requires strong infrastructure improvements across the grid, renewable sites, electric vehicle facilities, and rural access points to ensure a smooth transition.

6.4 Regulatory and Policy Constraints

Many government rules and policies create obstacles to India's energy shift goals. Without a single national energy policy, India faces investment challenges because of unclear rules for renewable energy projects. Different state rules work against national energy objectives and slow down project execution. The complicated permit process forms a major obstacle to developing renewable energy projects. The multiple steps needed to obtain approvals, along with land and environmental issues, create project setbacks and high costs, which make developers and investors hesitant. Today's rules do not effectively support both energy storage and grid connection systems. Old power grid regulations and limited encouragement for better energy storage systems block the effective linking of renewable energy sources like solar and wind to the power network. Current regulations create stability issues for renewable energy projects that supply electricity.

The government provides more financial help to fossil fuels than to renewable energy, making competition unfair. The current financial policies create unequal market conditions that reduce the appeal of switching to clean energy technologies. To solve these regulatory problems, India needs comprehensive policy changes, including easier approval standards and unified state and national energy strategies that enhance economic benefits for renewable projects. These steps are necessary to move India toward renewable energy faster and reach its climate objectives.

7. Energy Transition Opportunities

7.1 Electricity Sector Potential

India has many renewable energy opportunities that provide successful alternatives to fossil fuels. This section studies each part of renewable energy to show what effect it can have on India's future power system. As a nation with plentiful natural assets, India leads the way toward cleaner energy systems. Renewable energy development responds to environmental needs while also advancing economic benefits and strengthening India's energy supply system. India plans to use multiple renewable energy systems, including solar power plants, wind turbines, and biomass facilities. Each power source needs specific strategies because it brings different benefits and challenges. Due to the wide range of natural conditions across the country, different regions can take advantage of solar and wind energy production effectively.

The change from one energy form to another creates fresh opportunities across multiple sectors. The electric grid has great potential through renewable energy sources and system development. Construction companies gain advantages from sustainable building standards and combine renewable power with energy-saving measures. Using electric vehicles and improving public transit while decarbonizing logistics will provide new business opportunities for the transportation industry. The industrial sector aims to use low-carbon solutions while integrating renewable power sources and improving energy performance. Our report outlines each renewable energy type by analyzing how far each technology has progressed, its potential for expansion, and current plans for utilization. Our findings explain how India can use sustainable power sources to build a long-lasting energy system.

7.1.1 Solar Energy

The growth of solar energy in India will strongly drive the country's sustainable development because it holds great potential. The nation experiences about 300 days of sunshine each year, which makes solar energy an ideal energy solution. The National Solar Mission, launched in 2010, plans to build 100 GW of solar capacity by 2022 using both power grid links and rooftop systems. The growth of solar energy depends heavily on financial factors. Technological development and mass production have lowered PV panel prices, making solar energy an affordable power source compared to traditional energy. Government financial support from both central and state authorities attracts public-private investment in solar energy projects.

The sector faces issues with land rights and heavy startup costs, while linking to electricity networks remains difficult. India must implement supportive measures alongside financial and technology upgrades to access its large solar power capacity.

7.1.2 Wind Energy

Wind power in India represents a major opportunity to add new types of energy to the national supply system. In 2021, India stood in fourth position among global wind power producers, thanks to its installed facilities generating over 50 GW. India benefits from its long coastlines and mountainous regions when developing wind energy production. Tamil Nadu, Gujarat, and Karnataka lead the nation in wind power generation. MNRE plans to expand India's wind power infrastructure to 60 GW before 2022 with strong organizational assistance. Wind energy installations need significant upfront funding yet deliver affordable operations and sustainable power generation over time. New wind turbine technology with increased size and height brings better energy production and lower expenses.

However, the sector faces several challenges. Unpredictable wind patterns create energy generation problems that require better ways to store energy for later use. Building large wind farms faces community pushback as well as problems with both government approval and acquiring required land. For wind energy to keep developing successfully in India, it needs proper grid network support and fair government policies. When government agencies work with business partners, they can unlock wind energy's full potential to power India's renewable energy future.

7.1.3 Biomass and Other Alternatives

Grown biomass products as energy help farmers earn extra income while solving agricultural waste problems in rural areas. Biomass energy technology ranges from basic heat production through burning to complex bio-refining methods that produce both biofuels and biochemical products. The successful use of biomass technology appears in plants that turn animal and organic waste into energy. India also has untapped energy sources such as geothermal power and high-potential small hydro energy plants. Though India's use of geothermal energy remains at an early stage, the stored heat within the Earth provides a dependable energy supply. Small hydro projects utilize India's river and stream systems to produce local energy while remaining environmentally friendly.

Our investigation into these energy solutions identifies main problems such as expensive startup costs, technical difficulties, and the necessity of strong government policies. Using these energy sources in India would help to strengthen energy security and lessen environmental damage while making their power supply more varied.

7.2 Buildings Sector Potential

The building sector determines most of India's energy use and carbon emissions. However, the building industry contains powerful opportunities for advancing green development and moving to sustainable energy sources. The building sector in India must lead changes to power our transition toward low-carbon energy.

7.2.1 Energy Efficiency

Our biggest chance today lies in saving more energy. Many residential and commercial buildings in India waste energy needlessly. Upgrading how buildings retain heat and light energy, plus switching to better systems for lighting and heating, helps buildings use less energy. Property owners save money while society reduces its use of energy and produces less carbon pollution. The building sector requires government support through financial relief and technical assistance, plus building regulations that enforce energy efficiency minimums.

7.2.2 Renewable Energy Integration

Renewable energy adoption creates an important opportunity for the building industry. Solar panels on building roofs have gained widespread acceptance in India and could add much more to their total presence. Rooftop solar panels help building owners create renewable power and sell surplus electricity back to the grid through meter connections. Research could lead to better energy alternatives in the building industry by studying solar energy systems and using biomass or geothermal sources for heating.

7.2.3 Green Buildings

Standards from green building certification organizations set basic principles for better sustainable architectural practices. Certifications, including LEED and IGBC, guide builders by defining standards for energy-efficient designs, water-saving solutions, eco-friendly materials, improved air quality plans, and responsible waste disposal. Developers and property owners increase building health and sustainability through eco-friendly construction practices. Properties with these environmental certifications increase their market value while attracting people who care about eco-friendly living.

7.2.4 Electrification of Buildings

Many people are embracing building electrification as a promising new direction. Moving from using fossil fuels to power heating and cooking can decrease carbon emissions. Building electrification targets low-carbon living through electric heat pumps, induction cooktops, and electric vehicles. Building electrification succeeds only when clean electricity remains affordable and governments help create the necessary infrastructure.

7.2.5 Energy Storage

The building industry uses energy storage devices to help generate more power while saving it along the way. Battery storage equipment helps buildings store solar energy that exceeds their regular operational demands and supports the power grid. Renewable power systems can work more efficiently when energy storage devices help control peak demand and support the overall grid.

7.3 Transport Sector Potential

India's transport industry uses a large percentage of total energy and adds substantially to the nation's greenhouse gas emissions. The transport sector offers plenty of space to shift how energy is produced while pushing forward economic development. Switching to cleaner transportation systems is necessary to fight climate change while making the air safer and reducing energy dependency.

7.3.1 Electric Vehicles

India's transportation change begins with electric vehicles (EVs). The government has set ambitious targets for EV usage, and business expansion remains rapid. Driving EVs helps India benefit from cleaner air while saving money and improving the environment. The government provides discounts for buyers and upgrades EV charging facilities to encourage faster adoption. EV market growth comes from better batteries and lower costs, combined with rising public awareness about electric vehicles.

7.3.2 Public Transportation

Public transport helps control traffic congestion while lowering environmental impact. Building up our public transport network with more metro rail, buses, and suburban trains helps us achieve sustainable transportation. More organizations are choosing electric buses because they provide a clean and efficient way to move people. Making better use of current public transport systems with advanced technology and route planning decreases both energy use and pollution.

Shared mobility is becoming more popular in India. By sharing vehicles, more people can drive less, which lowers traffic problems and reduces air pollution. Shared transportation helps people use vehicles more effectively and decreases the demand for personal car ownership.

7.3.3 Charging Infrastructure Development

A solid charging system must be in place for electric vehicles to achieve widespread use. A complete charging system, both at public locations and in homes, needs to be built to ease EV driving concerns and expand EV ownership. Private companies and government institutions support charging network development, but this effort needs further improvement.

7.3.4 Non-motorized Transport

Making walking and cycling transportation options available helps build sustainable cities and promote good health. Setting aside space for walkers and bikers, plus enhancing road safety, will encourage people to use eco-friendly ways to get around.

7.3.5 Logistics and Freight

The logistics and freight industry offers many possibilities for changing how we use energy. Commercial vehicles can operate better on electricity or hybrid power to reduce emissions and save fuel. By rearranging delivery routes, combining shipments, and using advanced transportation methods, companies can reduce their energy usage while making the planet healthier.

7.4 Industrial Sector Potential

As India's economy grows, its industrial businesses create wealth and produce an outsized share of the country's environmental pollution. The industrial sector must move to low-carbon operations to meet India's climate targets and advance economic progress. This business area holds many opportunities to switch to new forms of power.

7.4.1 Energy Efficiency

Effective use of energy stands as the primary element for industrial emission reduction. Upgrades in technology, plus changes to production methods and heat recovery systems, will both decrease energy use and lower operating expenses for companies. By embracing IIoT and AI, industrial facilities can use modern tools to better manage their energy use while identifying ways to enhance performance. Government programs that help businesses save energy take several forms, including financial backing and energy assessment services.

7.4.2 Renewable Energy Integration

Renewable energy systems give industrial facilities a strong development opportunity. Solar power and biomass can produce renewable electricity in industrial locations. Our dependency on the central electricity supply decreases as we lower our energy expenses. To decrease their carbon impact, industries can purchase renewable energy certificates (RECs). Government support programs for renewable power generation help industrial companies choose clean energy sources.

7.4.3 Decarbonizing Industrial Processes

Moving industrial operations toward cleaner energy needs focused attention within the energy transition process. The main operations of steel-making and chemical plants require fossil fuels to run. Using low-carbon technologies and alternative raw materials helps decrease emissions. Steel producers need to test hydrogen technologies, while cement makers must develop CCS systems and new binding materials.

7.4.4 Circular Economy

The circular economy helps industries lower their emissions by reducing material use and waste. By using circular economy methods like material recycling and product renewal, industries can decrease their environmental impact.

8. Role of Government and Private Sector

Both the government and private sector need to work together to achieve India's energy transformation. Both parties help drive progress toward renewable energy and sustainability. The government leads this agenda by creating policy frameworks while offering subsidies and market advantages to promote renewable energy investments. It builds interlocking rules to control how businesses operate and manage market risks while creating reliable, long-term policies. Success in building more renewable energy systems depends largely on bringing private companies into the process. Changing energy systems requires joint efforts between government authorities and business partners. The government's authority covers policy creation, incentive setup, and long-term target setting. Organizations invest in new ideas while improving workflows to support energy transformation. When the government and private sectors collaborate on major infrastructure projects, they combine public funds with private operational expertise to decrease market risks.

In contrast, private enterprises bring innovation along with financial investments and run operations more effectively. Private energy companies help develop better technologies to make renewable energy more affordable and efficient. Market competition between private sector participants decreases energy costs and increases industry innovation. The private sector works with companies around the world to bring in leading energy transition technologies and successful approaches. Private companies can support national energy priorities better while protecting both the energy supply and decreasing their environmental impact. India needs both national government support and private partnerships to tackle the many challenges of transitioning its energy system. The next sections explore how policies impact private investments and develop those plans further.

8.1 Policy Interventions

Effective policies drive India toward a better energy future. The Indian government uses different policies to support renewable power adoption and decrease fossil fuel usage. The National Solar Mission serves as the main driver to help India become a solar energy world leader by deploying 100 GW of solar power technology by 2022. These solar projects support the National Action Plan on Climate Change, which aims to include climate measures in India's overall policy system. Recent official support for using solar and wind energy together demonstrates the government's commitment to protecting power grids and creating efficient land usage. The Renewable Purchase Obligation requirements make states purchase power from renewable sources to grow market demand for clean energy.

The government created financial and tax benefits aimed at boosting renewable energy initiatives. It provides accelerated tax benefits, Viability Gap Funding, and

GST tax cuts for renewable power components to help project developers reduce investment expenses. State governments implement special renewable energy rules to help their regions produce more energy. Central and state policy clarity, along with strict regulatory standards, keeps investors motivated to help India achieve its clean energy goals.

8.2 Private Sector Investments

Private sector investments lead the way in transforming India's energy system, as India moves from burning fossil fuels to generating renewable power. Private investors provide essential funding as well as bring fresh methods needed to expand renewable power projects. Businesses keep investing in solar and wind energy projects thanks to governmental policies that help, along with falling renewable technology prices and future financial gains. Major investments from Tata Power and Adani Green Energy now exist in solar power plants and wind turbine locations across many Indian states.

Private equity and venture capital companies invest more heavily in renewable energy businesses now. These financial backers want to join the industry growth and help achieve sustainable development goals. External investments boost project development and create better ways to perform operations while helping advance technology. International investors have become essential partners in support of these initiatives. This momentum is reinforced by direct investments from foreign firms, including Soft-Bank and Brookfield, in India's renewable energy projects. Foreign investors enable India to acquire advanced energy transition methods while propelling its shift to clean sources of power. Despite progress, financial risks and uncertain policies continue to be barriers for investors. Private companies and government institutions must partner to overcome these challenges on their way to a successful renewable energy system.

8.3 Public-Private Partnerships

Public-private partnerships (PPPs) serve as fundamental tools for India to reach its energy transformation goals. By working together with private partners, public sector entities use their resources to develop better energy solutions. PPPs help speed up infrastructure construction by sharing resources and developing solutions to financial and regulatory problems collaboratively. Both public and private sectors benefit from collaborations that allow the government to create supportive policies and incentives, while private companies deliver technological upgrades. PPPs have made important progress possible within India's national energy transformation programs. When local and private entities team up to set up renewable energy facilities, they work with government organizations to achieve these goals. These collaborations help manage major projects and reduce financial market volatility risks during technical system integration.

When private and public organizations work together, they can upgrade power lines to support renewable power production. The grid requires better lines, smarter control systems, and additional storage to function effectively. By combining government and business resources, the public and private sectors can build stronger power grids and increase efficiency in their operations. Partnering with public and private organizations helps build energy systems through joint research and technology education. When the private sector joins public sector activities, it brings modern industry technologies and business methods that help both parties build better renewable energy programs.

9. Case Studies and Success Stories

The progress toward clean power in India is evident through several successful energy projects that demonstrate how renewable systems work in practice. These results help guide future energy transitions while showing what effective methods look like in practice. Gujarat's state solar policy has become one of the top achievements in renewable energy development. In 2009, Gujarat established its Gujarat Solar Policy to encourage solar power production in state facilities. Through supportive regulations and payment benefits, Gujarat built Asia's largest solar park, which can generate 1,000 megawatts of power. The project added significant renewable energy capacity for the entire country while inspiring other states to follow suit.

The state of Tamil Nadu provides an example of success in building wind energy projects. Tamil Nadu leads with 29% of all wind power capacity in India because of its favorable wind resources and supportive government policies. By building large wind farms and connecting them to the state power grid, the region receives steady and renewable power, which reduces its fossil fuel consumption. Remote villages in Himachal Pradesh benefit from micro-hydro energy systems that show decentralized clean energy can work effectively. Projects providing electric power services have improved community development and living conditions in Indian villages.

9.1 State-Level Initiatives

As states begin leading India's energy transformation, they create important learning experiences about regional policy updates and energy rollouts. Indian states apply different strategies to adopt renewable power because their local features affect their energy plans. The state of Gujarat uses its high solar radiation to reach noticeable achievements. The state leads solar power production because its strong policies support both large solar park development and rooftop solar installation promotion. The state's approach to energy distribution helps communities across the region take part in improving their energy security.

The state of Tamil Nadu achieves strong results with its wind energy projects because of its favorable wind energy patterns. The state supports private sector wind power investments by creating easy regulations and offering favorable price deals. Despite being one state, Tamil Nadu generates a quarter of India's total wind power output. The wind and solar power programs in Karnataka show other states how to roll out renewable energy effectively. The state employs various positive energy policies, such as promoting market competition and securing power deals, while enhancing power grid network performance. Karnataka has built stronger energy infrastructure through these projects, allowing it to provide energy more reliably.

States that design specific energy plans achieve better results in attracting renewable energy use. Local energy projects make powerful contributions to the countrywide energy system, which works best when designed to fit local advantages and issues.

9.2 Successful Pilot Projects

Through practical tests, India confirms that using renewable energy works well for its power systems. The Gujarat Solar Park stands out as the main renewable energy development. Recognized globally as the largest solar farm, it produces over 1,000 MW of electricity. The project made Gujarat a solar energy leader, attracting both domestic and foreign investors while demonstrating how to develop larger solar plants. In Andhra Pradesh, the Wind-Solar Hybrid Project demonstrates best practices in energy generation. Developed by Greenko Group, the combination of wind and solar energy delivers dependable power for customers. The hybrid approach solves power quality problems from renewable power plants while saving space and construction resources.

In Odisha, the Chhotkei Microgrid Project demonstrates a proven approach for serving communities at the local level. By integrating solar panels and battery storage into one system, it supplies power to distant villages. The project proves renewable energy installations can transform rural energy scenarios while showing how distributed energy systems connect urban and rural communities. The Cochin International Airport became the first airport globally to use only solar energy for power. The 40 MW arrays show how to include renewable energy in major infrastructure development while reducing carbon pollution. Multiple trial programs show renewable energy works well in different environments, so they should be implemented across the country.

10. Discussion

The above analysis underlines that the energy transition in India is a multifaceted process shaped by financial, infrastructural, economic, and technological dimensions. India has made significant progress in enhancing the capacity of renewable energy sources by investing in policies like the Green Hydrogen Mission and production-linked incentive schemes. The change from fossil fuels toward sustainable energy sources has become essential for India. Despite being one of the fastest-developing nations, India requires an immediate solution for growing energy needs combined with a shift to cleaner sources of energy. Our power generation methods heavily depend on fossil fuels, especially coal, which provides 49% of our 441 GW total generating capacity. Renewable power sources, especially solar and wind energy, are quickly becoming more important components of our energy mix. Multiple forces bring about changes in our energy systems. The rising prices of foreign fuel supplies and the falling prices of solar and wind technologies determine how strongly economic factors push energy change. The need to fight global warming and clean up the air environment drives energy transition decisions. The advancement of solar PV innovation, combined with better energy storage systems, lets us move from fossil fuels to renewable power more easily at an economical rate.

Although India has opportunities ahead, it still struggles with multiple energy transition obstacles. High startup costs and investors' risk perceptions make it hard to finance renewable power projects today. The path to renewable energy adoption becomes more difficult when power lines need upgrading and government officials must acquire land. The energy transition meets strong resistance from government rules and practices.

Beyond its challenges, the energy transition offers many chances to develop different industries and shows positive growth for India's future. Solar and wind renewable energy hold substantial development potential within the electricity sector. Organizations in the building industry can develop improved energy efficiency programs and introduce renewable power sources. Transportation now includes more electric vehicles and introduces better public transit options. The industrial sector needs to make both energy efficiency gains and transitions to renewable energy technologies. These options lead directly to better economic performance and new employment opportunities while supporting environmental sustainability.

11. Conclusion and Future Outlook

The journey toward a new energy system in India has major benefits and tough barriers to overcome. Understanding all three aspects controls our success in building a sustainable future. Coal, oil, and natural gas power most of India's energy system today as its primary fuel sources. Our government now supports renewable power more than ever before. The urgent need to protect the environment and the planet, plus international requirements, drives, enables, or creates these opportunities. Various elements push India toward switching its energy system. Economic needs strongly drive decisions about energy transition because these changes can lead to new jobs while making energy sources safer and communities stronger. Environmental initiatives that reduce gas emissions and protect nature are becoming more important. New technology helps organizations deliver renewable energy solutions that perform better while also saving money.

Although we face challenges during the energy transition, our journey ahead still looks promising. Hard financial limits, along with building power networks, face difficult challenges, combined with government rules and policies that block progress. To make progress, our energy system needs effective financial plans backed by infrastructure projects and system-wide policy changes. India's shift to renewable energy faces a difficult path but promises better results in the future. As India strives for both economic development and sustainability, new pathways and market trends will direct its energy transformation path.

Economic Growth and Investments: India's expected economic growth provides ideal conditions for investing in its power systems. The government's push for renewable energy will attract both Indian and foreign investors. Through the International Solar Alliance and state-backed incentive programs, India stands out as a favorable market for renewable energy investment.

Technological Innovations: More advanced technology will make renewable energy options work better and become cheaper to use. Solar panel performance upgrades, combined with modern storage systems and intelligent grid platforms, will make renewable energy more affordable and reliable. Using artificial intelligence and big data analytics improves how energy moves through the power grid and helps people use less energy.

Policy and Regulatory Landscape: As India follows international climate commitments, new laws to help reduce carbon emissions will take effect. Regulations that reward cleaner technologies will help the country move away from providing tax breaks for fossil fuel production. Continued government support is essential to keep the energy transformation moving forward.

Social and Environmental Impact: Converting energy resources from fossil fuels helps create jobs and reduce environmental effects, bringing many advantages for society and the economy. India can strengthen its power supply by relying less on foreign fuel purchases, which benefits both the economy and the environment.

Through systematic planning and effective policies, India can move toward more sustainable energy usage with new technology. Several groups must work together to make India's energy transition succeed. Both government efforts and business actions determine success in this matter. Through policies and regulations, the government helps establish future objectives and motivates industry development. Private industry members invest funds while introducing new technology and improving their operations. Government and private companies team up for large infrastructure projects while managing market risks more effectively. Collaboration is essential to meet the urgent task of transforming energy systems. India's movement from current energy forms is crucial for building a better future. The plan provides benefits through stronger economic growth and energy security combined with job creation. For these advantages to materialize, detailed planning supported by strong policies and the further development of energy technologies is necessary to address current challenges. The outcome of India's shift to sustainable energy will define both the country's growth and resilience while advancing the global approach toward climate change.

References

Ang, T. Z., Salem, M., Kamarol, M., Das, H. S., Nazari, M. A. and Prabaharan, N. 2022. A comprehensive study of renewable energy sources: Classifications, challenges, and suggestions. Energy Strategy Reviews 43: 100939.

AP News. 2023, January 23. Clean energy gains a foothold in India, but coal still rules. https://apnews.com/article/3868417d77263e2f7977fe0d98fb1353.

Debanjan, M. and Karuna, K. 2022. An overview of the renewable energy scenario in India and its impact on grid inertia and frequency response. Renewable and Sustainable Energy Reviews 168: 112842.

Debnath, R., Mittal, V. and Jindal, A. 2022. A review of challenges from increasing renewable generation in the Indian Power Sector: Way forward for Electricity (Amendment) Bill 2020. International Journal of Sustainable Energy 41(5): 495–511.

Ding, Y., Mallapragada, D. and Stoner, R. J. 2024. The role of coal plant retrofitting strategies in developing India's net-zero power system: A data-driven sub-national analysis. arXiv.

Gupta, P. and Raut, M. 2021. Transmission of solar energy using fiber-optic. International Journal of Engineering and Technical Research 2278: 0181.
https://www.climatepolicyinitiative.org/
https://www.iea.org/
IEA. 2021, February 27. Renewable Energy Integration: India's Next Big Challenge. IEEFA.
IEA. 2021. Renewables Integration in India [Report]. IEA.
IFRI. 2025, February 20. Unlocking India's Energy Transition: Addressing Grid Flexibility Challenges and Solutions. IFRI Memo.
IIC (Impact Investors Council). 2024. Capital for Climate Bulletin: India's Energy Evolution.
Mohan, A., Sengupta, S., Vaishnav, P., Tongia, R., Ahmed, A. and Azevedo, I. L. 2021. Sustained cost declines in solar PV and battery storage are needed to eliminate coal generation in India. arXiv.
Paul, J. S., Sivan, A. P. and Balachandran, K. 2013, December. Energy sector in India: Challenges and solutions. pp. 519–524. In 2013 International Conference on Green Computing, Communication and Conservation of Energy (ICGCE). IEEE.
Reuters. 2025, February 3. Comment: India's renewables push holds lessons for the world.
Tongia, R. 2022. Balancing India's 2030 electricity grid needs management of time granularity and uncertainty: Insights from a parametric model. arXiv.

Chapter 4

Energy Decarbonization through Best Feasible Option for India

Yash Pal[1] and *Narendra N. Dalei*[2,*]

1. Background

India is now the 4th largest economy in the world (after the USA, China, and Germany), having overtaken Japan. India has experienced rapid industrialization, notably boosting its transportation and urbanization sectors since the New Industrial Policy in 1991. The rise in economic activity has culminated in a substantial increase in energy utilization. Consequently, this sector plays an essential role in the country's economic progress and provides the framework for its advancement. In accordance with several sources, India relies heavily on fossil fuels and coal, both of which contribute considerably to CO_2 emissions. The excessive use of such fossil fuels is a major concern for environmental sustainability. Over-reliance on fossil fuels exacerbates climate change by depleting ecosystems and increasing greenhouse gas emissions. India must switch to renewable energy sources to ensure a sustainable future, minimize its carbon footprint, and foster a more resilient, robust ecosystem.

1.1 India's Energy Sector

India's economy has grown tremendously over the last ten years, and the energy industry has become crucial to that development. India, after China and the U.S.,

[1] Research Scholar, Department of Economics, Central University of Himachal Pradesh, Sapt Sindhu Parisar-I, Dehra, Dist. Kangra, Himachal Pradesh (India)-177101.

[2] Professor, Department of Economic Studies, Central University of Punjab, Bathinda, VPO- Ghuddaq, Distt.- Bathinda, Punjab (INDIA)- 151401.

Email: yash.shalog@gmail.com

* Corresponding author: nndalei@cup.edu.in

DOI: 10.1201/9781003475385-4

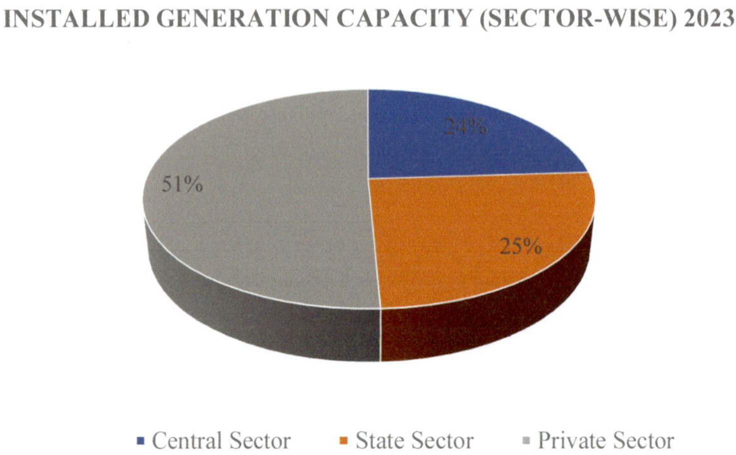

Figure 1. Sector-wise Installed Capacity (Source: Ministry of Power, GOI).

is the third-largest nation in the world in terms of primary energy consumption and power generation. The requirement for energy has become crucial for maintaining economic growth, as it drives up the demand for power as the economy grows. Energy production is classified into three categories: Central, State, and Private sectors. Each sector makes a substantial contribution to overall energy production. The sector-wise installed generation capacity is shown in the following graph.

Figure 1 illustrates sector-wise contribution of installed generation capacity, which reveals that the private sector accounts for 51%, followed by the state sector at 25.7% and the central sector at 24.3%.

Figure 2 depicts the installed power generation capacity. The figure reveals that 56.80% of the installed generation capacity is derived from fossil fuels, with coal constituting the most prevalent source at 49.10% of the total capacity, and 43% from non-fossil fuels. Renewable energy sources comprise 41.40%, with solar power accounting approximately 16.10% and wind 10.30%. Additionally, hydropower comprises 11.20%, while nuclear energy contributes 1.60% of the total capacity. Fossil fuels, mainly coal—which accounts for approximately half of the total installed capacity—remain the major source of energy for India. Nevertheless, non-fossil fuel sources, such as renewables and nuclear, account for a substantial proportion (43.00%) of the capacity, indicating a considerable transition toward cleaner energy sources.

The two forms of energy demonstrated in the above figure are renewable energy and conventional energy, often known as non-renewable energy (Figure 3). It also shows that non-renewable sources of energy, such as natural gas and coal, release CO_2 and are finite in nature. Overexploitation of these natural resources jeopardizes environmental sustainability. On the other hand, wind and solar power, being two renewable sources of energy that are free from emissions and abundantly available, encourage sustainability.

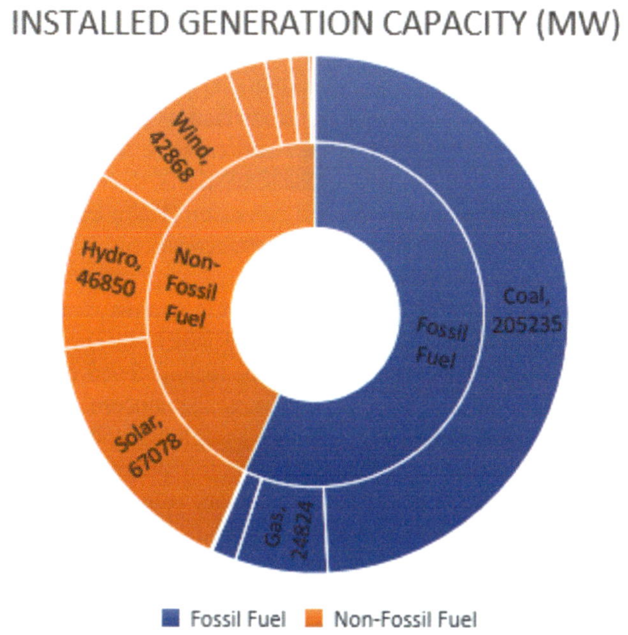

Figure 2. Installed Generation Capacity (FUELWISE). Source: Compiled by authors from the website of Ministry of Power, GOI.

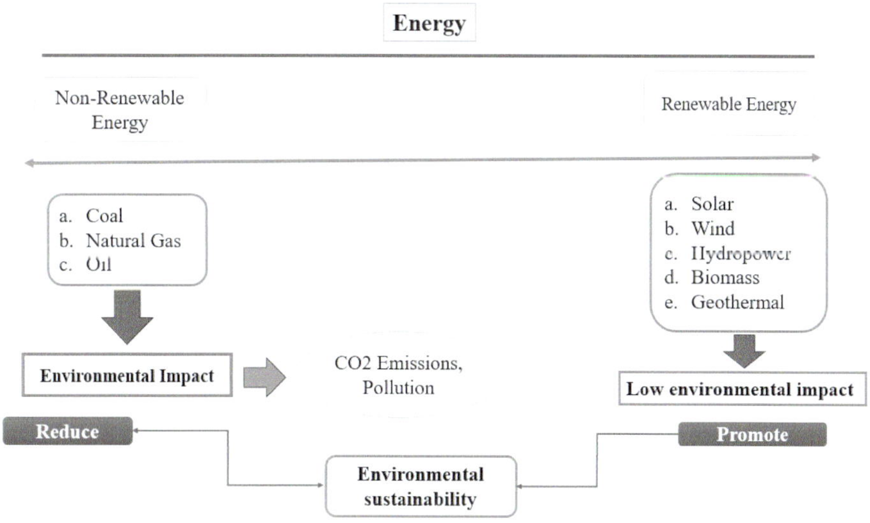

Figure 3. Type of Energy; Source: Author Compilation.

1.2 India's Perspective on Climate Change

A serious threat to our planet is none other than climate change that demands serious and urgent collaborative effort worldwide to mitigate its impact. However, the global reaction to this subject has been seen as unsatisfactory and has led to substantial condemnation (Sridharan 2023). Developed nations have traditionally contributed considerably to climate change and releases of GHG through industrialization, urbanization, transportation, and economic expansion (Kabir et al. 2023). Figure 4 shows that both developed and developing countries consistently generate greenhouse gas emissions from various sectors.

Despite having the capability to win the battle against climate change, they have been incompetent in doing so successfully (Suri 2023). Furthermore, they have failed to keep their commitments to provide technical and financial support to underdeveloped nations in their attempts to combat the concern.

Throughout the decade since its emergence as a global leader, India has become more concerned about climate change. According to (Sharma 2023), India, widely recognized as the representative of the South Asian nations, is a key actor in the worldwide dialogue on climate change. Its attitude toward climate change has significant ramifications for global attempts to reduce greenhouse gas emissions. India is the seventh most susceptible nation to climate change. Nevertheless, India has taken steps to reconcile the detrimental impacts of climate change with its unique mitigation and adaptation requirements domestically. *India's annual CO_2 emissions constitute only 3% of the global total, with a per capita emission of 1.9 tonnes—significantly lower than most developed nations (Panigrahi 2024). Furthermore, India continues to progress and develop climate change mitigation frameworks with the objective of producing a net-zero emission economy.*

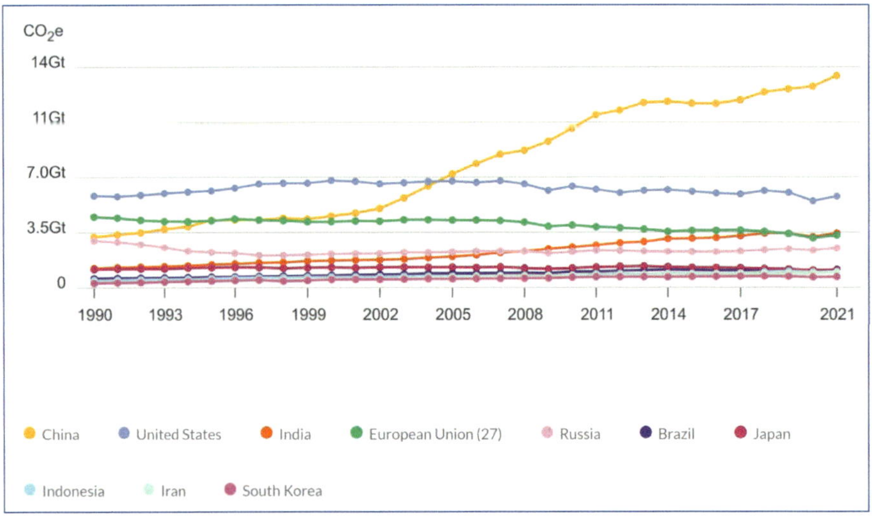

Figure 4. Top Global Nations of Greenhouse Gas Emissions including (Industrial, Energy, Agriculture, and Waste sector) from 1990–2021; Source: Climate Watch.

The COP-28 was held in Dubai, UAE, from 30th Nov. to 12th Dec. 2023. Delegates from 197 nations attended to explain their actions to combat global warming and address future climate change concerns. One of the summit's noteworthy features was that it was the very first time each delegation officially acknowledged its accomplishments under the Paris Agreement of 2015. India also played a key role in the summit. In the presidential session, India's Prime Minister was scheduled to speak on "Transforming Climate Finance." The most important objective was to concentrate on strengthening climate finance and enhancing its accessibility and affordability for emerging economies. One of the foremost announcements made by the Prime Minister was, "*India is one of the few economies in the world that is on track to meet the Nationally Determined Contribution (NDC) targets. We have already achieved non-fossil fuel targets related to emissions intensity 11 years ago. We have achieved the non-fossil fuel target nine years ahead of schedule. And India has not stopped at this. Our target is to reduce emissions intensity by 45 percent by 2030. We have decided that we will increase the share of non-fossil fuel to 50 percent.*"

Furthermore, India's commitment to promoting net-zero emission frameworks, addressing issues pertaining to climate change, and its participation in the G20 on environmental issues demonstrate the country's burgeoning geopolitical significance. When India inherited the G20 chair in 2023, its presidency's theme was "One Earth, One Family, One Future." The theme emphasized India's conviction in the interconnectedness of human existence across the world. Consequently, addressing societal concerns requires a collaborative worldwide commitment. India's standpoint on climate change is straightforward, with a strong commitment to mitigation. However, being one of the fastest-growing economies in the world, India remains the third-highest emitter of CO_2 after the United States and China. *It has undertaken several initiatives to reduce its environmental footprint.* The energy transition of India requires large-scale exploitation of solar and wind energy. Green finance and investment in green technology to sustain sustainable development are the major focus of India. The National Action Plan on Climate Change and the International Solar Alliance (ISA) highlight India's proactive stance. The ultimate objective of these initiatives is to accomplish net-zero carbon emissions and decarbonization. By stimulating innovation and global collaboration, India is setting the path for a more sustainable future that balances economic expansion with safeguarding the environment.

1.3 International Solar Alliance (ISA)

The ISA, founded in 2015 by India and France at the COP21 conference in Paris, advocates for solar energy as a sustainable energy option for climate change mitigation. As India's first international organization, it promotes global solar collaboration, energy security, and the transition to greener energy systems, with 120 member and signatory countries (Press Information Bureau 2024). Furthermore, it is a platform that aims to promote solar energy use and collaboration among governments, organizations, financial institutions, and the private sector to develop a unified strategy for addressing energy issues and climate change. The 7th Session of the ISA (November 3 to 6, 2024) held in New Delhi focused on accelerating the implementation of solar energy in member countries, particularly in areas with limited electricity access. The ISA initiated several programs to encourage solar energy projects and promote worldwide collaboration. These included the SolarX Startup Challenge, STAR-C Initiative, Global Solar Facility, Viability Gap Funding Scheme, Solar Data Portal, International Solar Festival, Green Hydrogen Innovation Centre, ISA Knowledge Series, and World Solar Reports (Ministry of New and Renewable Energy 2024). These initiatives aimed to accelerate solar energy implementation and promote global collaboration. The ISA also explored the synergies between solar energy and hydrogen, establishing itself as a prominent proponent of solar energy globally. Similarly, ISA's third High-Level Conference on New Technologies for Clean Energy Transition, in partnership with India's Ministry of New and Renewable Energy, the Asian Development Bank, and other collaborators, focused on advanced solar technology, storage options, and solar's potential for socio-economic and environmental development. India, as a founding member and a key driver of the ISA, plays a crucial role in spearheading global efforts to reduce emissions and mitigate climate change. Furthermore, it provides a pathway for the transition toward clean energy solutions. India and the ISA are promoting a low-carbon future through capacity development, financial innovation, and affordable solar technologies, demonstrating their commitment to the Paris Climate Agreement and fostering collective initiatives to revolutionize the global energy landscape.

2. Decarbonization

Decarbonization is the process of minimizing the amount of carbon dioxide (CO_2) emissions from human activities produced through the combustion of fossil fuels. It aims to mitigate climate change by switching to low- or zero-carbon energy sources and implementing sustainable practices. It seeks to improve energy efficiency and substitute fossil fuel-derived energy with zero-emission renewable energy wherever possible (Hubler and Loschel 2013). Key initiatives include expanding the adoption of wind and solar energy, promoting energy efficiency, electrified transportation, and installing carbon capture and storage (CCS) systems (Figure 5).

2.1 India and Decarbonization Initiatives

As a global leader, India is highly committed to mitigating climate change. Various initiatives, targets, and goals to actively support the decarbonization policy have been initiated by India. India is a climate leader with a strong commitment to renewable energy transformation. At the COP26 meeting in Glasgow, India's Prime Minister proposed an initiative known as *Panchamrit* or the Gift of Five Elixirs (CEA 2020). Achieving 500 GW of non-fossil energy capacity, generating 50% of its energy from renewable sources, reducing overall estimated carbon emissions by one billion tons, and cutting carbon intensity by less than 45% are the four major goals India aims to achieve by 2030. Furthermore, India is committed to accomplishing net-zero carbon emissions by 2070.

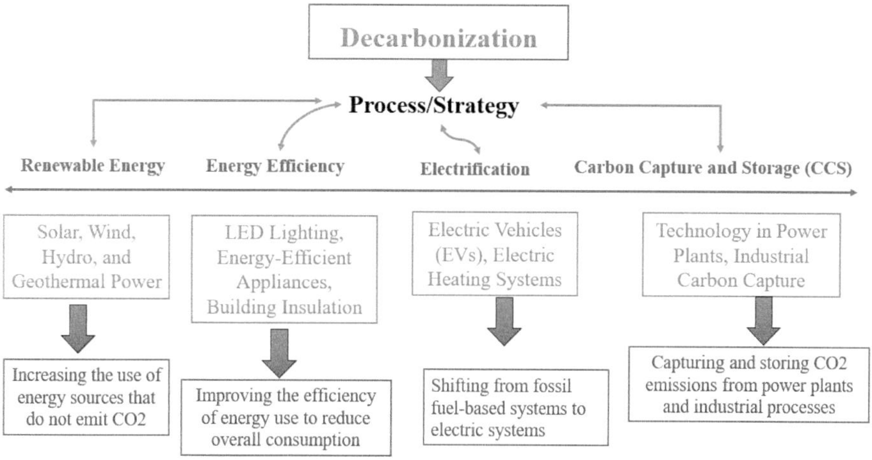

Figure 5. Strategies for Decarbonization; Source: Compilation from Literature.

At the COP21 held in Paris in 2015, India announced its Nationally Determined Contributions (NDCs) (Figure 6). The year 2015 marked a turning point in India's transition to a low-carbon electricity sector. During this period, India intended to increase its non-fossil fuel power-generating capacity to 40% by 2030 and to install 175 GW of renewable energy sources by 2022. Furthermore, India has reduced emissions by 28% over 2005 levels and targets to accomplish its NDC goals by 2030 (Shankar et al. 2022).

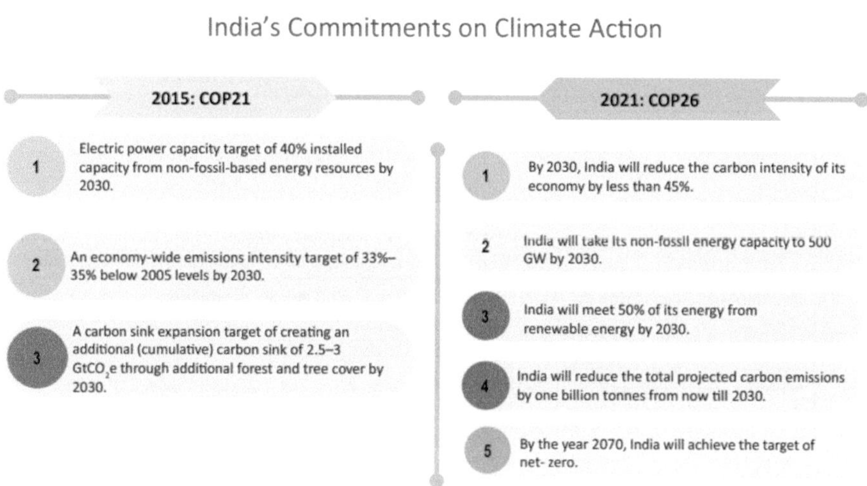

Figure 6. Sustainable Goals Targets; Source: Shankar et al. (2022).

These are the key commitments, targets, and goals for decarbonization and sustainable development. Additionally, the table below contains information about several initiatives that encourage sustainability and decarbonization:

Table 1. Major initiatives for sustainable development.

Sr.	Initiatives	Year	Description
1	Coalition for Disaster Resilient Infrastructure (CDRI)	2019	CDRI Objectives: • Construct resilient infrastructure. • Enhance existing structures. • Minimize infrastructure losses from natural disasters.
2	Infrastructure for Resilient Island States (IRIS)	2021	IRIS Collaboration with SIDS • Identifying technical collaborations for infrastructure system upgrades. • Enhancing resilience capabilities. • IRIS's Contribution to SIDS Accelerated Modalities of Action. • Strengthening infrastructure resilience to climate change and disaster risks.
3	One Sun One World One Grid (OSOWOG)	2021	Global Collaboration Program • Fosters trust among participating nations through political participation. • Enhances collaboration between international financial and technical institutions. • Focuses on knowledge and information exchange among regulators and grid operators through peer-to-peer learning networks.
4	Global Biofuels Alliance		Alliance Aims: • Enhance sustainable biofuel acceptance. • Advancement of biofuel technology. • Increasing consumption of sustainable biofuels. • Address ILUC concerns.
5	National Green Hydrogen Mission	2022	Global Green Hydrogen Hub • Production, utilization, and export of green hydrogen. • Aims to reduce carbon emissions. • Enhances energy security. • Creates sustainable economic growth opportunities.
6	International Solar Alliance (ISA)	2015	• Boost renewable energy production • Promote adaptation of solar energy and increase advance energy technology
7	Mission LiFE – Lifestyle for the Environment	2023	• Aims to mobilize one billion Indians and global citizens for environmental conservation. • Promotes mindful, deliberate use of resources to shift from "use and dispose of" mentality. • Aims to form "Pro-Planet People" (P3), committed to eco-friendly lifestyles.

Source: Panigrahi (2024)

Table 1 illustrates numerous initiatives that India leads and promotes at the global level. The primary goals of these initiatives are to promote sustainable development and mitigate the effects of climate change. Furthermore, the ultimate purpose of those initiatives is to increase decarbonization.

2.2 *Sustainable Transformation: The Feasible Energy Path to Decarbonization*

India has significant potential for decarbonization through various solutions, as Figure 2 illustrates. One viable alternative is shifting from conventional energy to renewable sources like solar and wind. Solar and wind energy have significant potential for output, accounting for around 30% of total renewable energy generation.

2.3 *Renewable Energy Expansion*

2.3.1 Solar Energy

Solar energy is the most prominent source of energy in the present day. This type of energy is generated through nuclear fusion in the sun, which means it is a clean energy source. Solar energy happens to be the most widely and extensively growing source of energy across the globe. It is the most efficient approach for reducing carbon emissions in India's power grid. With plentiful sunlight throughout every season, India has an immense opportunity for solar power generation. Expanding the solar power grid may substantially decrease dependency on fossil fuels, leading to a decrease in emissions. Initiatives such as the National Solar Mission are intended to increase solar power production, rendering it a key component of India's renewable energy plan. Photovoltaic farms, rooftop installations, and novel innovations such as solar-powered water pumps could all contribute to creating long-term growth while also providing energy security and ecological benefits. Expanding solar energy utilization is critical to meeting India's decarbonization ambitions.

Figure 7 demonstrates that solar energy output has increased rapidly throughout the years. Electricity generation grew dramatically starting in 2020, reflecting that India is making significant efforts to shift to a low-carbon future.

Figure 8 demonstrates the anticipated potential for solar energy generation in India's different states and union territories in 2021–22. Rajasthan (142,310 MW) has the highest solar energy potential in India, followed by Jammu and Kashmir (111,050 MW), Maharashtra (64,320 MW), and Madhya Pradesh (61,660 MW). Furthermore, the overall potential for solar energy production in India totals 748,990 MW. Similarly, according to the Ministry of New and Renewable Energy, the National Institute of Solar Energy states that in 2023, the potential solar power is estimated at approximately 750 GWp.

The findings illustrate a steady rise in solar energy production throughout the years, with substantial potential for additional expansion. Government milestone initiatives are intended to accelerate the attainment of ambitious targets in this sector. This energy transformation is critical to minimizing emissions and accelerating decarbonization efforts. It is imperative that the government increase funding and support at the household level for the adoption of solar energy.

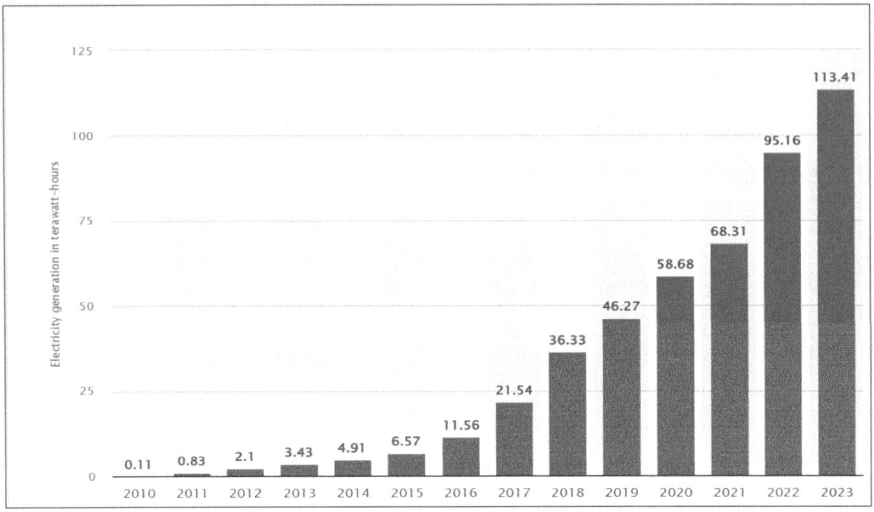

Figure 7. Trends of Solar Energy Generation 2010-2023 Source:https://www.statista.com/statistics/1401414/india-solar-power generation/#:~:text=Solar%20power%20generation%20in%20India,of%20280%20gigawatts%20by%202030.

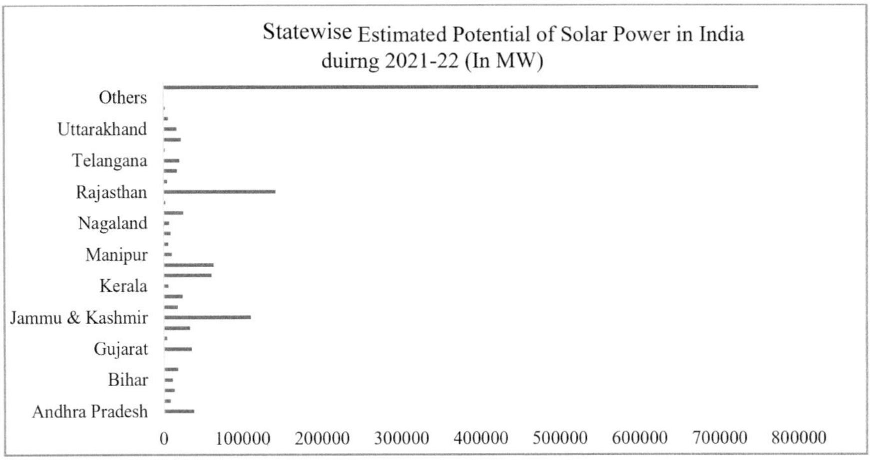

Figure 8. State-wise Estimated Potential of Solar Energy Generation; Source: Ministry of New and Renewable Energy.

2.3.2 Wind Energy

Wind power is another one of the most achievable solutions for India's energy transition and reducing sustainability challenges, and it has immense potential to accelerate the nation's transformation to a more sustainable future. India has the perfect location to efficiently utilize wind energy due to its massive coastline zones and vast rural landscapes. Furthermore, in the long run, wind power installation has proven to be cost-effective and makes a substantial contribution to decarbonization by minimizing dependency on fossil fuels. Wind energy is becoming a more feasible preference for sustainable development as technological improvements minimize costs and raise the effectiveness of turbines. By investing in wind power plants, India may accomplish its energy transition targets, sustain its energy sovereignty, and minimize the consequences of climate change.

Since the mid-1990s, private sector investments have driven tremendous growth in India's wind industry. The upsurge was encouraged by an assortment of fiscal and non-fiscal incentives. Furthermore, beyond growing domestic market demand, India has also created a globally competitive wind power manufacturing business (Annual report 2022–23, Ministry of New and Renewable Energy).

Figure 9 shows a wind energy estimate for 2029. It demonstrates that wind power generation increases with the period of time, with a projected 96.58 billion kilowatt-hours by 2029. This also reflects India's rapid attainment of energy transformation targets.

Figure 10 illustrates the installed capacity of grid-linked wind power in India throughout different time periods. It shows the amount of wind energy capacity added each year (in MW) and the total cumulative wind power capacity in specific years. Between 2016–17, India witnessed a massive increase in wind power capacity, with 7,192 MW added during a single year. Furthermore, total wind power capacity has continuously expanded over time.

Figure 11 illustrates the projected wind power potential at a 120-meter height for various Indian states and union territories (UTs) in 2021–22. Gujarat, Karnataka, and Rajasthan have significant wind power potential, with Gujarat having the highest capacity of 1,42,560 MW, while Karnataka and Rajasthan have capacities of 1,24,155 MW and 1,27,756 MW, respectively. The overall wind power potential across all states and union territories in India is 6,95,508 MW, exhibiting considerable potential for wind energy expansion across the country.

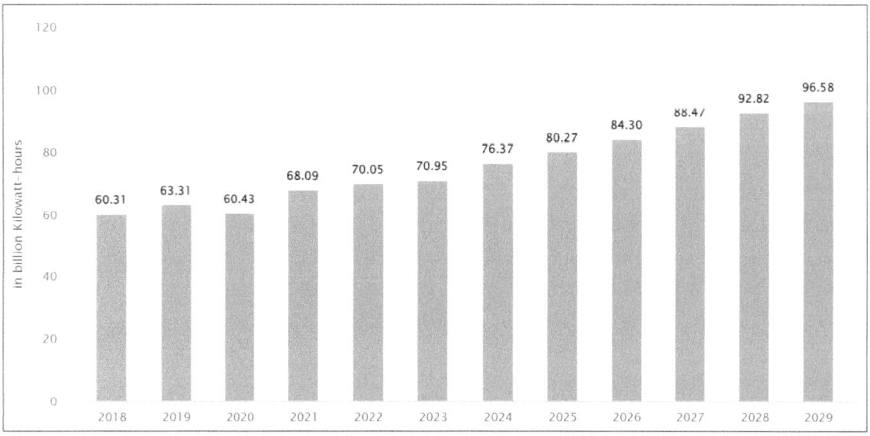

Figure 9. Trends of Wind Energy Generation 2018–2029. Source: https://www.statista.com/outlook/io/energy/renewable-energy/wind-energy/india#production.

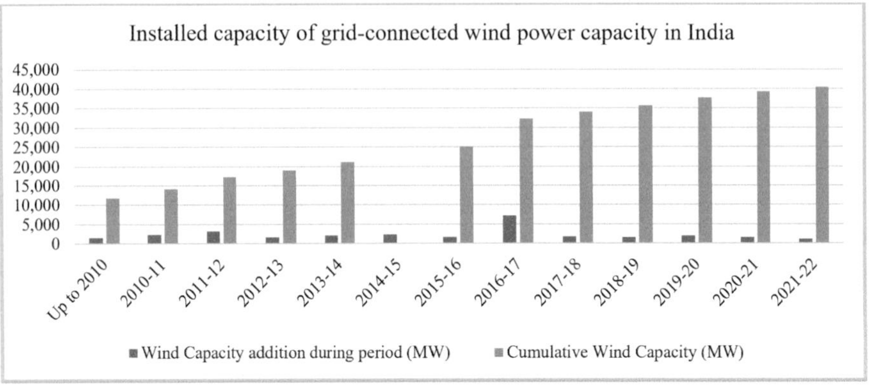

Figure 10. Installed capacity of grid-linked wind power; Source: MNRE, GoI.

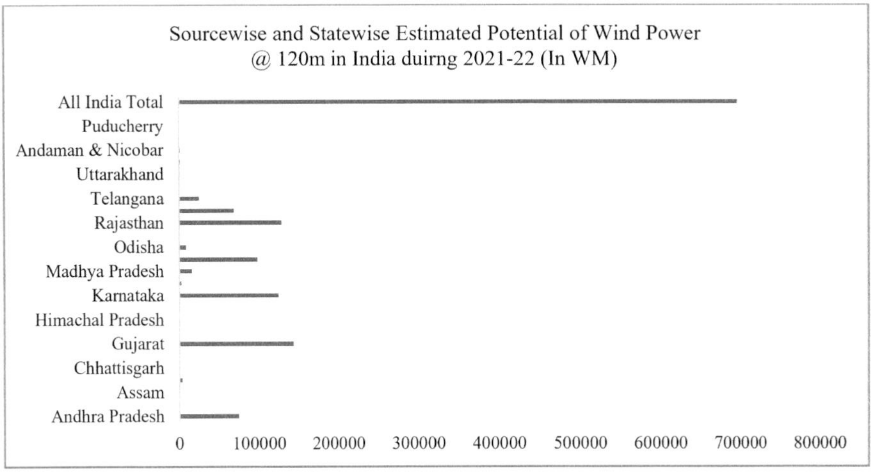

Figure 11. Estimated Potential of Wind Power for Various States; Source: MNRE, GoI.

The statistics illustrate that India has immense opportunity for energy transformation based on its projected wind power capacity. Such tremendous potential emphasizes the viability of attaining the country's decarbonization goals and strengthening environmental sustainability. Alongside significant contributions from states such as Gujarat, Karnataka, and Rajasthan, wind power emerges as a crucial substitute for minimizing carbon emissions and cleaning up India's energy networks. Through the utilization of this renewable resource, India may take significant steps toward a more sustainable future and establish wind energy as an essential component of its energy policy.

2.3.3 Hydropower, Biomass and Bioenergy

Hydropower is an important clean energy source in India, which constitutes around 13% of the total installed power capacity. It has an immense capacity of roughly

145,000 MW, spread across several locations, exploiting huge river systems. India's energy mix includes both large-scale hydropower projects and small and micro hydropower units. The Government of India intends to significantly increase the utilization of hydropower for a sustainable supply of green energy. Thus, fossil fuel energy dependence will be reduced substantially, supporting the decarbonization of the power sector. Furthermore, biomass and bioenergy are vital in India's renewable energy sector, contributing around 4%. Derived from forestry waste, agro-residues, and organic solid waste from municipalities, bioenergy projects promote efficient waste utilization, waste management, and greenhouse gas reduction. This supports rural development and job creation, contributing to sustainable development goals.

2.3.4 Hydropower Capacity

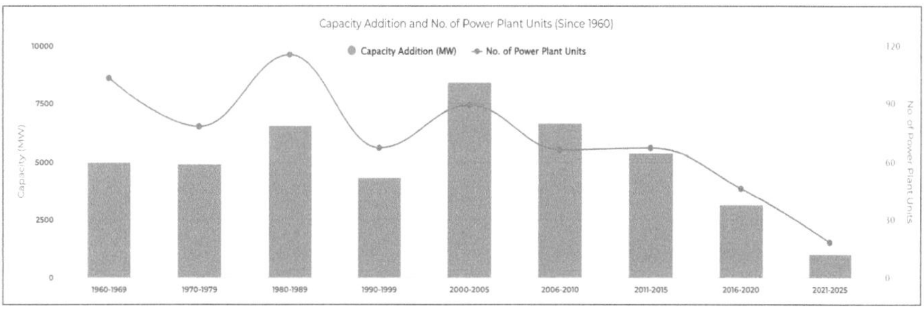

Figure 12. Trends of Hydropower Capacity. Source: NITI Aayog.

2.3.5 Bioenergy

Figures 12–13 reveal that hydropower and biofuels have risen substantially over time. These figures also show that the energy transition aim will be met within a particular time period. The increasing use of hydropower and biofuels helps to ensure environmental sustainability by lowering emissions and enhancing cleaner energy generation.

Figure 14 indicates that Karnataka, Himachal Pradesh, and Arunachal Pradesh are the top producers of small hydropower. The immense rise in hydropower assists in improving environmental sustainability. These renewable sources help alleviate the detrimental effects of climate change by lowering emissions and strengthening cleaner energy generation. The tremendous potential of hydropower assures the future's sustainability by promoting decarbonization efforts and ensuring a smooth energy transition.

The earlier overview illustrates that clean energy is the most effective approach for accomplishing decarbonization targets and ensuring environmental sustainability. India has enormous potential for renewable energy generation, as highlighted in the above figures. Such potential, along with the rapid rise of renewable energy, will make a substantial contribution toward attaining the Sustainable Development Goals (SDGs) and net-zero carbon emissions.

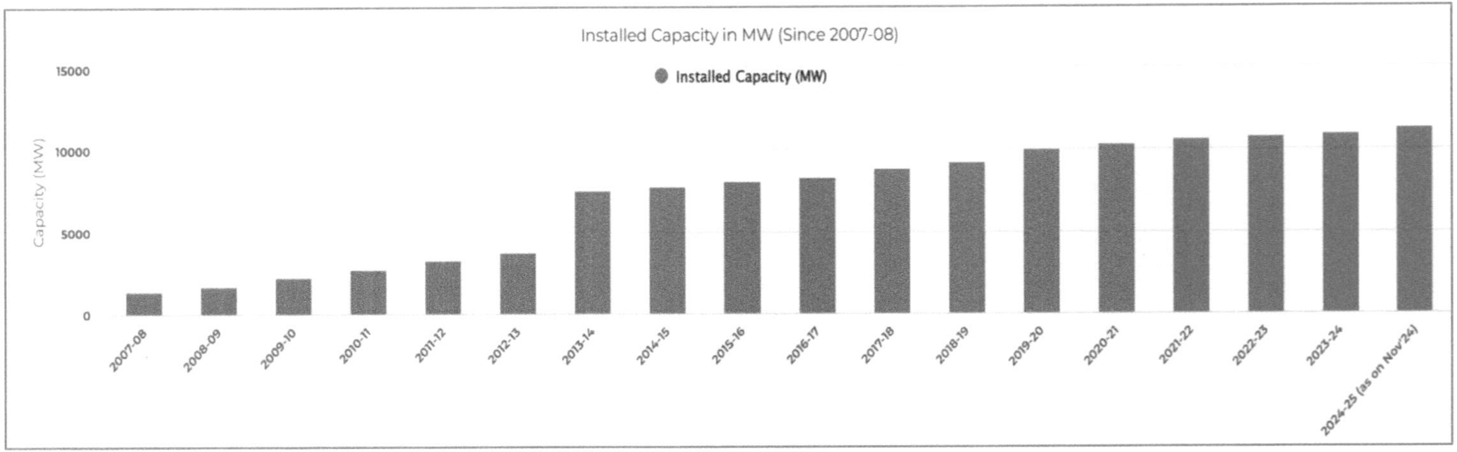

Figure 13. Trends of Bioenergy Capacity. Source: NITI Aayog.

S.No.	State	Total Potential		Project Installed						Project under implementation	
				upto 2022-23		upto 2023-24		Total			
		No.	Total Capacity MW	Nos.	Capacity MW	Nos.	Capacity MW	Nos.	Capacity MW	Nos.	Capacity MW
1	Andhra Pradesh	359	409.32	45	163.31	0	0	45	163.31	1	1.20
2	Arunachal Pradesh	800	2064.92	157	133.11	0	0	157	133.11	5	6.05
3	Assam	106	201.99	6	34.11	0	0	6	34.11	0	0
4	Bihar	139	526.98	29	70.7	0	0	29	70.7	0	0
5	Chhattisgarh	199	1098.2	10	76	0	0	10	76	0	0
6	Goa	7	4.7	1	0.05	0	0	1	0.05	0	0
7	Gujarat	292	201.97	22	91.64	0	0	22	91.64	2	21.66
8	Haryana	33	107.4	9	73.5	0	0	9	73.5	0	0
9	Himachal Pradesh	1049	3460.34	202	969.71	0	0	202	969.71	45	251.84
10	UT of Jammu & Kashmir	103	1311.79	20	146.68	4	23.25	24	169.93	5	29.65
11	UT of Ladakh	199	395.65	30	40.99	1	2.00	31	42.99	5	5.6
12	Jharkhand	121	227.96	6	4.05	0	0	6	4.05	0	0
13	Karnataka	618	3726.49	170	1280.73	0	0	170	1280.73	6	16.45
14	Kerala	238	647.15	39	266.52	2	10	41	276.52	4	32.85
15	Madhya Pradesh	299	820.44	14	123.71	0	0	14	123.71	3	7.7
16	Maharashtra	270	786.46	72	381.08	1	1.20	73	382.28	7	8.10
17	Manipur	110	99.95	8	5.45	0	0	8	5.45	0	0
18	Meghalaya	97	230.05	5	32.53	1	22.50	6	55.03	1	3.0
19	Mizoram	72	168.9	20	45.47	0	0	20	45.47	0	0
20	Nagaland	98	182.18	14	32.67	0	0	14	32.67	1	2.4
21	Odisha	220	286.22	13	115.63	0	0	13	115.63	3	56.5
22	Punjab	375	578.28	59	176.1	0	0	59	176.1	5	4.05
23	Rajasthan	64	51.67	10	23.85	0	0	10	23.85	0	0
24	Sikkim	88	266.64	18	55.11	0	0	18	55.11	0	0
25	Tamil Nadu	191	604.46	21	123.05	0	0	21	123.05	0	0
26	Telangana	94	102.25	30	90.87	0	0	30	90.87	0	0
27	Tripura	13	46.86	3	16.01	0	0	3	16.01	0	0
28	A&N Islands	7	7.27	1	5.25	0	0	1	5.25	0	0
29	Uttar Pradesh	251	460.75	10	49.1	0	0	10	49.1	1	1.5
30	Uttarakhand	442	1664.31	103	218.82	0	0	103	218.82	6	65.55
31	West Bengal	179	392.06	24	98.5	0	0	24	98.5	0	0
	Total	**7133**	**21133.61**	**1171**	**4944.30**	**9**	**58.95**	**1180**	**5003.25**	**100**	**514.10**

Figure 14. Potential sites, installed projects, and on-going projects in Small Hydro Power Sector. Source: https://cdnbbsr.s3waas.gov.in/s3716e1b8c6cd17b771da77391355749f3/uploads/2024/10/202410141365502799.pdf.

Nevertheless, merely constructing or creating more capacity is insufficient. Efficient use of renewable energy sources is also critical. India's energy landscape requires essential attention on optimizing the usage of renewable energy to reduce wastage and boost effectiveness. Smart grids, energy storage technologies, and improved energy management systems may assist India in utilizing renewable energy more efficiently and consistently (Figure 15).

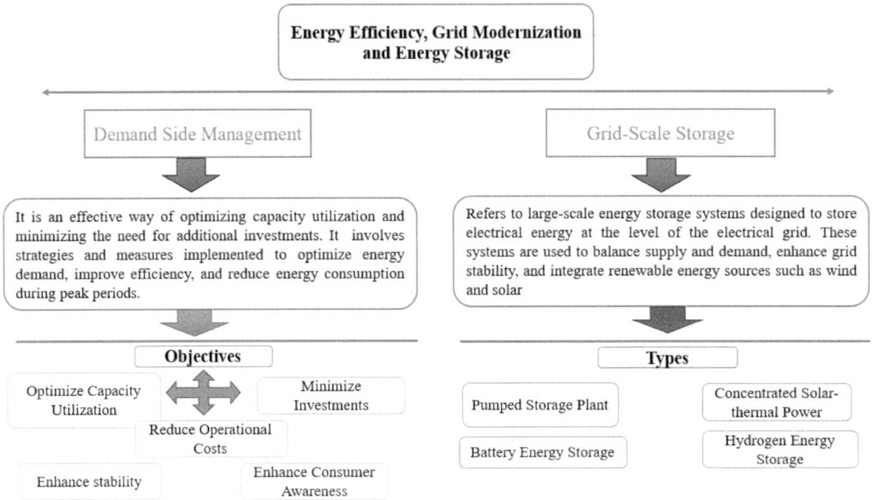

Figure 15. Energy Efficiency. Source: Author Compilation.

The storage grid as Pumped Storage Hydropower (PSP) facilities, which account for 94% of the worldwide stored-energy capacity, are extremely effective, attaining over 80% efficiency over a complete cycle and producing up to 10 hours of power (Kumar et al. 2022). The recently released draft National Electricity Policy (NEP) 2021 emphasizes the urgent necessity for the widespread implementation of Pumped Hydro Storage (PHS) to sustain the electrical grid (Shankar et al. 2022). Furthermore, CSP (Concentrated Solar Power) projects utilize reflectors that gather daylight and retain it in molten salt, which in turn powers a turbine to produce energy. This enables CSP to offer electricity on demand, making it an efficient renewable energy storage solution, and India has a noteworthy possibility for CSP, projected at over 2,700 GW (FOWIND 2018). Hydrogen-based energy storage has also been taken into consideration for long-term storage because it is technically better than battery storage. Green hydrogen, generated from renewable energy rather than fossil fuels, is vital for eventual decarbonization (Oliveira et al. 2021) and is becoming more affordable as the cost of renewable energy falls. Finally, it reflects that India has immense potential in clean energy production and also has a high possibility for storage in the future. Additionally, incorporating renewable energy into various sectors, including transportation, industry, and domestic areas, would increase its influence, contributing to a more sustainable and efficient energy system. To summarize, while India's large renewable energy potential is an immense advantage, attaining decarbonization goals demands not just increased generation capacity but also effective usage. By accomplishing this, India would be able to make significant progress toward its aspirational ecological goals and support worldwide efforts to mitigate climate change and enhance resilience.

3. Decarbonization Through CCSU Technology

Carbon Capture, Storage, and Utilization (CCSU) is another prominent decarbonization option. It serves to minimize CO_2 emissions and ameliorate the effects of climate change. CCUS refers to a variety of technologies targeted to capture CO_2 emissions from industrial activity sources like power stations, refineries, and other industrial sites, as well as recovering existing CO_2 from the environment, which can then be stored in a safe place and further converted into usable products (Li et al. 2013). Thus, it is a group of technologies for capturing CO_2 emissions from large sources such as electrical power plants and industrial sites. There are basically three basic kinds of CO_2 capture technologies: physical, chemical, and biological techniques. Among the physical capture techniques, membrane-based CO_2 separation has gained increasing attention due to its energy efficiency and operational simplicity, wherein selective membranes allow preferential permeation of CO_2 from mixed gas streams, as illustrated in Figure 16 (Ji and Zhao 2017). Biological and chemical techniques frequently capture, store, or consume CO_2 in the same location. Physical capturing frameworks that absorb an enormous amount of CO_2 usually deposit it underground (Dalei and Joshi 2022).

These geologic structures include exhausted oil and gas reserves, unmineable coal deposits, deep saline reservoirs, and deep seas. The CO_2 accumulated in these locations has the potential to remain there for many years.

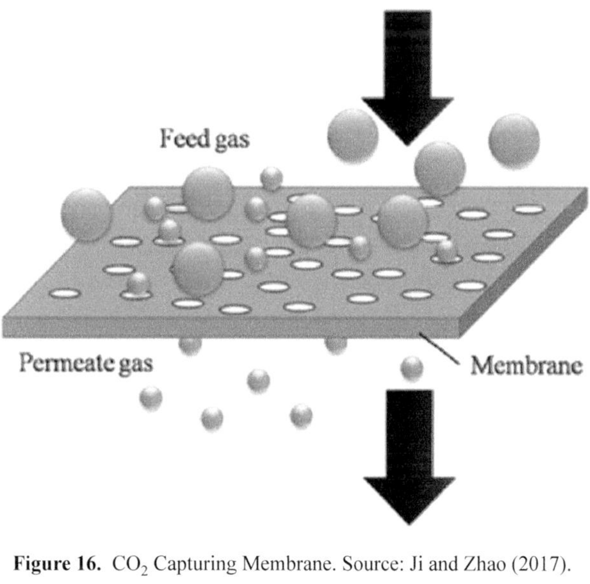

Figure 16. CO_2 Capturing Membrane. Source: Ji and Zhao (2017).

Table 2. Outline of Carbon Capture, Storage, and Utilization (CCSU) Technologies.

	Description	Details
Carbon Capture	Capture CO_2 emission from the source	**Pre-combustion capture method:** Entail transforming the fuel into a gas combination of hydrogen and CO_2 before burning it.
		Post-combustion capture: Once natural gas, coal, and other fuels are burned, carbon is extracted from the exhaust gasses.
		Oxy-fuel combustion: When fossil fuels are burnt in oxygen, flue gas mostly comprises CO_2 and water vapor.
Storage	Ensuring that collected CO_2 is stored securely and safely to circumvent atmospheric release.	Once CO_2 is captured, it is squashed into a fluid and transported via pipeline, vessel, commuter rail etc. Then, it is injected into bottomless geological layers, commonly at a depth of 1 km or more, and enduringly stowed in impoverished oil and gas basins, coalbeds, or deep saline aquifers.
Utilization	Using stored CO_2 in different applications.	Enhanced Oil Recovery (EOR) Industry sector Agriculture applications
Significance	The mechanism is vital for lowering the level of emissions in the atmosphere. Furthermore, it also boosts environmental sustainability and mitigates climate change consequences.	

Source: Dalei and Joshi (2022)

Table 2 demonstrates an overview of CCSU technology processes. The figure depicts one type of technology, but there are many more available, and they also have different suitable places to store it safely. Finally, there are several pathways for the optimum utilization of captured CO_2 across multiple sectors, including fuels, chemicals, food and beverages, enhanced oil recovery, and mineralization processes, as illustrated in Figure 17 (Ho et al. 2019).

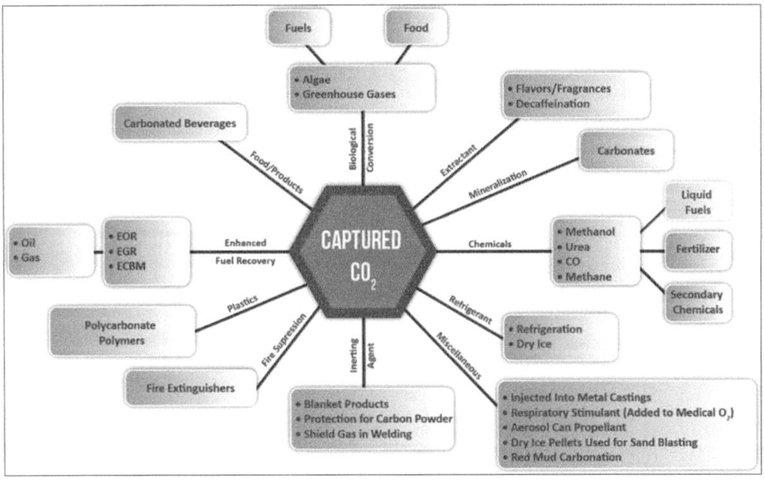

Figure 17. Utilization of Captured CO_2 Emission; Source: Ho et al. (2019).

CCSU is a commonly employed technique for minimizing CO_2 emissions around the world. It is noteworthy as it permits both the storage and application of CO_2 in different sectors, as illustrated in the figure. Additionally, it is a novel approach to promoting decarbonization and environmental sustainability. Nevertheless, it also has disadvantages, especially the possibility of leaks and deterioration, necessitating precautionary measures and free space. Additionally, over time it becomes cost-effective with technological advancements. Subsequently, the question arises: which is preferable—the CCSU mechanism or energy transition? Our subsequent debate will seek to identify which is more efficient and feasible.

4. Energy Transition Vs. CCSU: A Feasible Option for Sustainable Future

As the economy expands, it is critical to mitigate the effects of climate change. To solve this issue, we have two key options: rely on renewable energy sources or use CCSU technology. Among these two options, which one is the best solution? For this, it requires an analysis of environmental impact, cost-effectiveness, durability, etc.

Sustainability and Environmental Impact: Renewable energy sources, including solar and wind, are naturally sustainable, releasing a small amount of emissions or contaminants during operation (Androniceanu and Sabie 2022). Environmental protection using green energies is gradually becoming an everyday occurrence. The replacement of conventional sources of energy would be highly beneficial in environmental, social, and economic aspects (Jaiswal et al. 2022). Furthermore, several pieces of literature have revealed that renewable energy plays a significant role in reducing carbon emissions and promoting sustainability (Saidi and Omri 2020, Hao et al. 2021, Panwar et al. 2011). In contrast, CCSU technologies have immense capacity to capture up to 90% of CO_2 emissions from industrial sources. Nevertheless, this approach has serious disadvantages. The entire process is extremely lengthy and energy-intensive, necessitating significant resources. Moreover, CCSU requires considerable space for storage and poses the possibility of leakage (Bose et al. 2024) and deterioration, which might result in serious disasters and long-term harm to human populations and the environment. This will pose a high cost globally. Therefore, we can conclude that it may not be considered the finest solution for mitigating CO_2 emissions. Furthermore, in terms of economic benefits, the costs of green energy technologies have been gradually declining (Jiang et al. 2021). The Draft National Electricity Plan 2022 projects that by 2031–32, the capital cost of solar energy plants will be Rs 53.3 million per MW, whereas wind power facilities will cost Rs 77.9 million per MW (Takyar 2023). Furthermore, Shankar et al. (2022) reported that advancements in technologies have brought reductions in the cost of solar and wind power.

Figure 18 demonstrates that the levelized costs of electricity for utility-scale solar and wind technologies between 2009 and 2019 have been reduced substantially. Similarly, there is further evidence reflecting that renewable energy has become cost-effective over time. CCSU projects require substantial expenditures due to the high expenses of CO_2 acquisition, transportation, and storage (Goel 2017). The

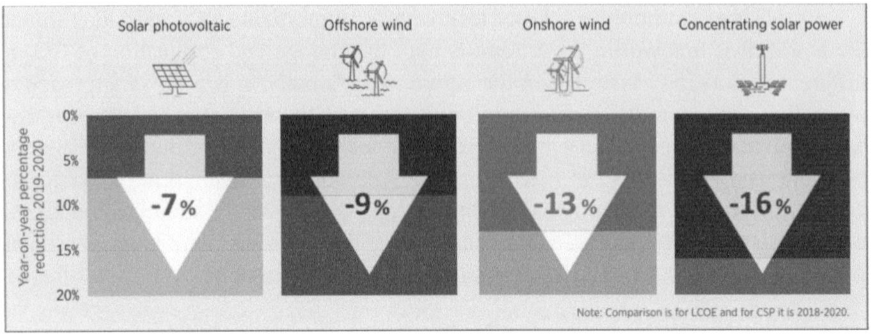

Figure 18. Percentage cost of solar and wind power reduction; Source: Shankar et al. (2022).

financial stability of CCSU is frequently reliant on significant government subsidies and institutional assistance. Additionally, technologies, including wind and solar, are substantially installed and affordable. Improvements in power Storage and network integration are improving reliability as well as effectiveness. In contrast, CCSU is still in progress, with just a few main projects in existence throughout the world. The technique must undergo substantial development before it can be considered scalable. Renewable energy generally enjoys broad public support due to its environmental benefits and potential to reduce energy costs over time, whereas CCSU technologies face public resistance because of concerns over CO_2 storage safety, long-term feasibility, and their effectiveness in addressing climate change.

5. Challenges of India's Fossil Fuel Dependency: Social, Economic, Technological and Regulatory Issues

India, a rapidly growing economy with a population of 1.4 billion, utilizes both conventional and renewable fuel sources in its power industry, including coal, oil, gas, solar, wind, biomass, industrial waste, and hydroelectric units. Furthermore, India, the third largest global energy user, is also experiencing a significant increase in energy consumption. Additionally, the nation's energy sector is primarily reliant on fossil fuels but is also enhancing its reliance on renewable and nuclear energy sources. The transition from coal to renewable energy will take decades, but India still needs to adapt to the fluctuating functioning of renewables. Despite this, India will continue to rely on natural gas and environmentally friendly fossil energy sources.

Furthermore, India's population is rapidly growing, leading to a surge in energy consumption. The consumption of fossil fuels like coal, oil, and gas generates heat energy and releases greenhouse gases that contribute to global warming. Renewable energy sources offer a solution to both energy and climate change. The economy, environment, and energy are interconnected, with a bi-directional causal relationship. Advancements in technology are needed to regulate and use electricity from alternative energy sources, addressing climate change by controlling carbon

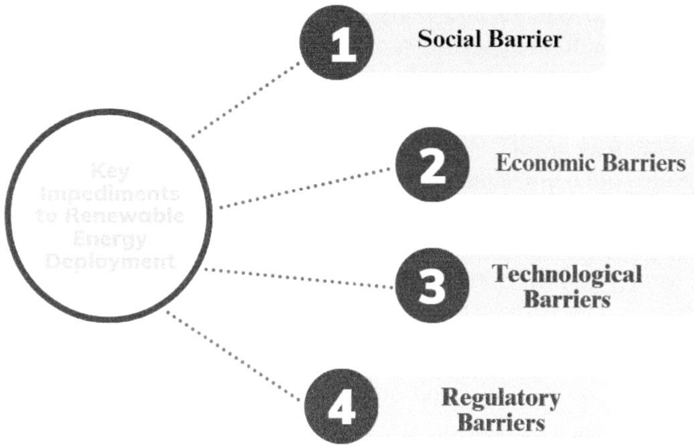

Figure 19. Key barriers affecting the application of renewable energy. Source: Luthra et al. (2015), Kumar et al. (2024a), Gopi et al. (2024).

emissions. Renewable energy promotes environmental sustainability and provides power to disadvantaged individuals in isolated regions without conventional electricity access, thereby protecting the planet. However, the widespread adoption of renewable energy technologies is constrained by multiple challenges, including social acceptance issues, economic and financial constraints, technological limitations, and regulatory and policy barriers, as illustrated in Figure 19. Moorthy et al. (2019) identified four primary barriers affecting the application of renewable energy: economic barriers, social barriers, regulatory barriers, and technological barriers.

5.1 Social Barrier

Sustainable development relies on socially acknowledged technical systems and appropriate policies and regulatory instruments (Luthra et al. 2015, Kumar et al. 2024b). Public awareness and information dissemination are hindered by inadequate information about ecological and financial advantages, a lack of proper understanding of renewable energy technologies, and reservations about renewable energy installation projects' financial viability (Aparisi-Cerda et al. 2024). Furthermore, insufficient availability of skilled professionals is another barrier to the widespread adoption of renewable energy. Incompetent technical workers and insufficient training institutions hinder the promotion of green energy projects. Therefore, there is a need for education in renewable energy courses and sufficient training to cultivate the skills necessary for the adoption of renewable energy projects (Moorthy et al. 2019).

5.2 Economic Barriers

Substantial initial capital, insufficient financial institutions, a scarcity of investors, competition from fossil fuels, and reduced subsidies compared to conventional fuels are all economic barriers to renewable energy adoption (Kumar et al. 2024). Furthermore, compared to renewable energy projects, fossil fuel projects are easier for developers to get funding and loans. There is a scarcity of financial mechanisms and institutions for funding green projects, making investments seem hazardous and dissuading investors. Renewable energy projects require substantial initial capital expenditure, and the net payback time is prolonged due to reduced efficiency (Umamaheswaran et al. 2024). Stringent lending policies limit access to finance, making ventures often unviable. Furthermore, immaterial expenses, such as the costs associated with environmental and societal harm, are not included in the overall cost of fuel. Therefore, understanding these effects is crucial for assessing the true expense of using fossil fuels for energy production.

5.3 Technological Barriers

Renewable energy implementation faces numerous technological obstacles, including inadequate infrastructure, inefficient operational and maintenance knowledge, insufficient research and development efforts, and technical complexities. Access to modern technology is limited, especially in developing regions, and countries that serve as barriers to renewable energy use are expensive to acquire (Gopi et al. 2024). Renewable energy power facilities often have supplementary transmission lines to link to the main grid, which need improvement or alteration. The nascent nature of renewable energy technology leads to a lack of understanding about its operation and maintenance, which is essential for efficiency. The unavailability of equipment, components, and spare parts requires significant manufacturing expenses, leading to increased costs and expenditure. Investment in R&D is inadequate to make renewable energy financially competitive with fossil fuels, and there are insufficient standards, processes, and guidelines for durability, dependability, and performance. Energy storage is a significant challenge, as the availability of solar or wind energy is not constant, and electrical systems must balance supply and demand (Jafarizadeh et al. 2024).

5.4 Regulatory Barriers

The expansion of renewable energy has been hindered by several factors, including the absence of effective national policies, bureaucratic and administrative obstacles, insufficient incentives, and a deficiency of standards (Anjanappa and Samant 2024). Robust regulatory regulations are essential for sustainable growth and addressing the disparity between renewable and non-renewable energy sources. However, the absence of appropriate policies leads to misunderstandings among departments and hinders the implementation of renewable energy projects. Insufficient fiscal incentives, such as feed-in tariffs, have led to elevated costs that impede the industry's development, operation, and maintenance. Administrative and bureaucratic intricacies, such as planning delays and prolonged permission processes, also contribute to the challenges faced in implementing renewable energy projects. A disparity exists between the policy objectives established by governments and the actual outcomes achieved via implementation. Governments must address these commitment issues by formulating policies that provide clear understanding of significant laws and regulatory matters, promoting the sector as stable and growth-oriented. Lastly, the absence of standards and certifications, which ensure equipment and components conform to importing company specifications, engenders misunderstandings and unnecessary challenges for energy producers.

6. Implications of the Transition to Renewable Energy: Policy, Infrastructure, and Future Outlook

Despite the consideration of various policies and global initiatives, the goal of decarbonization remains incomplete. Policies related to solar energy systems are already in place in India. However, the policies are not effectively implemented at the ground level. Furthermore, non-renewable energy utilities hold significant market power, making renewable energy sources like solar and wind difficult to enter. Start-ups face significant obstacles due to competition with established entities, infrastructure, and policies. To succeed, start-ups must demonstrate scalability and substantial energy output. Increased government investment in sustainable energy could create equitable conditions.

India's policy implications include promoting sustainable energy generation through legislative initiatives and regulatory structures, enhancing technical infrastructure, and developing more stringent rules to promote renewable energy adoption. To increase renewable energy consumption, the government must restructure the existing energy usage framework, which primarily relies on traditional coal sources. Additionally, advancing government investments in renewable energy initiatives is crucial for achieving ecological sustainability goals. The Indian government should advocate for the adoption of renewable energy such as solar and wind to reduce dependence on fossil fuels and create an environment favorable to human well-being. The country's large population and rapid growth have significant implications for conventional fuel use and greenhouse gas emissions. To alleviate these effects, the government should systematically replace conventional fossil fuels with sustainable and renewable energy sources. Environmental regulations should be enacted, including incentives for businesses to use innovative technology to reduce pollution. The government should focus on promoting the use of cleaner technology and enhancing productivity. Furthermore, India's policy strategy should utilize management techniques to tackle urgent ecological challenges through programs focused on renewable energy and research and development. Solar power solutions, wind power, and solar photovoltaic (PV) technology have significant potential for sustainable energy substitution. By reducing emissions and reaching carbon neutrality, India can achieve its environmental goals.

Furthermore, understanding the advantages of renewable energy facilities is crucial. The study suggests incorporating the comparative advantages of renewable energy sources and the disadvantages of fossil fuel utilization into educational materials to prepare the new generation for the energy crisis. Additionally, public-private partnerships for domestic energy generation are recommended, and the government should enhance public awareness about climate change and the potential benefits of domestic renewable energy facilities. Moreover, the government should implement subsidy programs and interest-free loan schemes to promote the installation of domestic energy systems. Eliminating or reducing import tariffs on renewable energy facilities can promote a sustainable economy, reduce fuel imports, and decrease CO_2 emissions.

7. Conclusion

Although green energy sources provide multiple advantages, there are still impediments to their widespread acceptance. These consist of substantial outlays in grid infrastructure, storage options to accommodate intermittent power, and legislative frameworks that promote market integration. While CCSU technologies can help eliminate emissions from preexisting fossil fuel infrastructure, transitioning to renewable energy sources is the most environmentally friendly, cost-effective, and socially acceptable choice for meeting long-term sustainability goals. Finally, taking into account all aspects of clean energy, it becomes apparent that India's generation and storage efficiency has grown throughout the years. Additionally, projections demonstrate that India has immense potential for enhanced power generation via solar and wind energy. Furthermore, advancements in technology have reduced the cost of production over time. Moreover, these energy sources encourage sustainable development, do not pose safety concerns or require vast open spaces, and definitely help in achieving net-zero carbon emissions. Finally, it is clear that renewable energy, such as solar and wind power, is the most viable solution for decarbonization.

References

Androniceanu, A. and Sabie, O. M. 2022. Overview of green energy as a real strategic option for sustainable development. Energies 15(22): 8573.

Anjanappa, J. and Samant, S. 2024. Exploring Financing Mechanisms for Scaling Clean Energy Solutions in India's Social Enterprises for Energy Poverty Alleviation. Available at SSRN 4762727.

Aparisi-Cerdá, I., Ribó-Pérez, D., García-Melón, M., D'Este, P. and Poveda-Bautista, R. 2024. Drivers and barriers to the adoption of decentralised renewable energy technologies: A multi-criteria decision analysis. Energy 305: 132264.

Bose, D., Bhattacharya, R., Kaur, T., Pandya, R., Sarkar, A., Ray, A. et al. 2024. Innovative approaches for carbon capture and storage as crucial measures for emission reduction within industrial sectors. Carbon Capture Science & Technology 12: 100238.

CEA. 2020. Optimal Generation Capacity Mix for 2029–30. Central Electricity Authority. Retrieved July 24, 2024, from

Climate Watch. (n.d.). Historical GHG emissions. Retrieved December 5, 2024, from https://www.climatewatchdata.org/ghg emissions?end_year=2021&regions=TOP&sectors=agriculture%2Cenergy%2Cindustrial-processes%2Cwaste&start_year=1990.

Dalei, N. N. and Joshi, J. 2022. Potential matching of carbon capture storage and utilization (CCSU) as enhanced oil recovery in perspective to Indian oil refineries. Clean Technologies and Environmental Policy 24(9): 2701–2717.

FOWIND. 2018. From Zero to Five GW: Offshore Wind Energy Outlook for Gujarat and Tamil Nadu (2018–2032). Global Wind Energy Council.

Goel, M. 2017. CO_2 capture and utilization for the energy industry: outlook for capability development to address climate change in India. Carbon Utilization: Applications for the Energy Industry 3–33.

Gopi, R., Pathak, D. V. and Pratap, S. 2024. Barriers and drivers for sustainable public transportation in Indian context. Green Energy and Intelligent Transportation 3(1): 100141.

Hao, L. N., Umar, M., Khan, Z. and Ali, W. 2021. Green growth and low carbon emission in G7 countries: how critical the network of environmental taxes, renewable energy and human capital is?. Science of the Total Environment 752: 141853.

Ho, H. J., Iizuka, A. and Shibata, E. 2019. Carbon capture and utilization technology without carbon dioxide purification and pressurization: a review on its necessity and available technologies. Industrial & Engineering Chemistry Research 58(21): 8941–8954.

https://cea.nic.in/old/reports/others/planning/irp/Optimal_mix_report_2029-30_FINAL.pdf.

Jafarizadeh, H., Yamini, E., Zolfaghari, S. M., Esmaeilion, F., Assad, M. E. H. and Soltani, M. 2024. Navigating challenges in large-scale renewable energy storage: Barriers, solutions, and innovations. Energy Reports 12: 2179–2192.

Jaiswal, K. K., Chowdhury, C. R., Yadav, D., Verma, R., Dutta, S., Jaiswal, K. S. et al. 2022. Renewable and sustainable clean energy development and impact on social, economic, and environmental health. Energy Nexus 7: 100118.

Ji, G. and Zhao, M. 2017. Membrane separation technology in carbon capture. Recent Advances in Carbon Capture and Storage 59–90.

Jiang, T., Ji, P., Shi, Y., Ye, Z. and Jin, Q. 2021. Efficiency assessment of green technology innovation of renewable energy enterprises in China: a dynamic data envelopment analysis considering undesirable output. Clean Technologies and Environmental Policy 23: 1509–1519.

Kabir, M., Habiba, U. E., Khan, W., Shah, A., Rahim, S., Patricio, R. et al. 2023. Climate change due to increasing concentration of carbon dioxide and its impacts on environment in 21st century; a mini review. Journal of King Saud University-Science 35(5): 102693.

Kumar, A., Pal, D., Kar, S. K., Mishra, S. K. and Bansal, R. 2022. An overview of wind energy development and policy initiatives in India. Clean Technologies and Environmental Policy 1–22.

Kumar, P., Kumar, P. and Aggarwal, V. 2024a. A behavioral study on villagers' adoption intention and carbon neutrality toward rooftop solar photovoltaic systems in India. International Journal of Energy Sector Management 18(1): 200–227.

Kumar, P., Gupta, S. and Dagar, V. 2024b. Sustainable energy development through non-residential rooftop solar photovoltaic adoption: Empirical evidence from India. Sustainable Development 32(1): 795–814.

Li, L., Zhao, N., Wei, W. and Sun, Y. 2013. A review of research progress on CO_2 capture, storage, and utilization in Chinese Academy of Sciences. Fuel 108: 112–130.

Luthra, S., Kumar, S., Garg, D. and Haleem, A. 2015. Barriers to renewable/sustainable energy technologies adoption: Indian perspective. Renewable and Sustainable Energy Reviews 41: 762–776.

Ministry of New and Renewable Energy. 2024, November 4. The International Solar Alliance hosts the seventh session of its annual assembly with representatives from 103 member & 17 signatory countries. Press Information Bureau. Retrieved from https://pib.gov.in/PressReleasePage.aspx?PRID=2070655.

Ministry of Power. 2023, June 12. Power sector at a glance: All India. Central Electricity Authority (CEA). Government of India. Retrieved December 5, 12, 2024 from https://powermin.gov.in/en/content/power-sector-glance-all-india.

Modi, N. 2023, December 1. Special Address by the Prime Minister at the inauguration of High-Level Segment of COP-28 of HoS/HoG. Press Information Bureau. Retrieved July 24, 2024, from https://pib.gov.in/PressReleasePage.aspx?PRID=1981612.

Moorthy, K., Patwa, N. and Gupta, Y. 2019. Breaking barriers in deployment of renewable energy. Heliyon 5(1).

NITI Aayog (https://iced.niti.gov.in/energy/electricity/generation/pipeline-capacity/bio-power).

NITI Aayog (https://iced.niti.gov.in/energy/electricity/generation/pipeline-capacity/hydro).

Oliveira, A. M., Beswick, R. R. and Yan, Y. 2021. A green hydrogen economy for a renewable energy society. Current Opinion in Chemical Engineering 33: 100701.

Panigrahi, V. 2024, 16 April. India Spearheading the Climate Change Mitigation Parade: A Focus on Its Global Initiatives. Indian Council of World Affairs. Retrieved July 24, 2024, from https://www.icwa.in/show_content.php?lang=1&level=1&ls_id=10754&lid=6831#:~:text=India's%20defining%20role%20in%20spearheading,better%20resources%20than%20other%20countries.

Panwar, N. L., Kaushik, S. C. and Kothari, S. 2011. Role of renewable energy sources in environmental protection: A review. Renewable and Sustainable Energy Reviews 15(3): 1513–1524.

Press Information Bureau. 2024, November 6. India shines bright! A new era in solar energy: Progress driven by the International Solar Alliance. Ministry of New & Renewable Energy. https://static.pib.gov.in/WriteReadData/specificdocs/documents/2024/nov/doc2024117431401.pdf.

Saidi, K. and Omri, A. 2020. The impact of renewable energy on carbon emissions and economic growth in 15 major renewable energy-consuming countries. Environmental Research 186: 109567.

Shankar, A., Saxena, A. K. and Idnani, T. 2022. Roadmap to India's 2030 Decarbonization Target. New Delhi: The Energy and Resources Institute.

Sharma, S. 2023, January 16. India's Climate Change Policy: Challenges and Recommendation. Indian School of Public Policy. Retrieved July 24, 2024, from https://www.ispp.org.in/indias-climate-change-policy-challenges-and-recommendations/.

Sridharan, S. 2023, September 16. Climate change: Global South suffers from developed world's inaction. Policy Circle. Retrieved July 24, 2024, from https://www.policycircle.org/opinion/climate-change-inaction-inertia/.

Suri, S. 2023, June 28. It's time for climate justice- A Global South perspective on the fight against the climate crisis. Retrieved July 24, 2024, from https://www.orfonline.org/research/a-global-south-perspective-on-the-fight-against-the-climate-crisis.

Takyar, Sarthak. 2023, April 25. Price Trends: Solar and wind power costs and tariffs. Retrieved September 12, 2024 from https://renewablewatch.in/2023/04/25/price-trends-solar-and-wind-power-costs-and-tariffs/.

Umamaheswaran, S., Dar, V., Prince, J. B. and Thangaraj, V. 2024. Risk perception as a barrier to renewable energy finance—a study of debt investors in the Indian context. International Journal of Energy Sector Management 18(6): 1511–1530.

Chapter 5

Application of Grey Multiple Criteria Decision Making in Sustainable Energy Optimization

Prachi Jain

1. Introduction

Energy sustainability is the topmost demand in today's world that must be balanced to achieve economic, environmental, and social objectives. The complexity of the problem is increased by the presence of conflicting multi-criteria and the inherent uncertainty and incomplete data for decision-making. Multi-Criteria Decision-Making (MCDM) methods are the best tools to evaluate and choose the most suitable energy sustainability alternatives. The integration of grey system theory with MCDM results in Grey Multi-Criteria Decision-Making (GMCDM), which has the ability to handle incomplete, uncertain, and ambiguous information. Grey MCDM methods are particularly suitable for energy sustainability problems where qualitative assessments and limited or uncertain data are involved. Two major techniques in the discussed area are Grey Analytic Hierarchy Process and Grey Entropy. Grey AHP extends the standard AHP by incorporating grey numbers to better represent uncertain pairwise comparisons, while Grey Entropy enables the objective weighting of criteria under uncertainty by measuring information entropy in grey environments.

This chapter presents the application of Grey AHP and Grey Entropy methods to energy sustainability evaluation through two case studies. The first case study

ABES Engineering College, Ghaziabad, India.
Email: prachi.jain.ime@gmail.com

DOI: 10.1201/9781003475385-5

employs Grey AHP for selecting the best alternative for a sustainable energy project. The second case study applies Grey Entropy to optimize a smart grid. Both case studies illustrate the effectiveness of Grey AHP and Grey Entropy in supporting robust, informed decision-making under uncertainty.

The subsequent section of the chapter provides the literature review and fundamentals of Grey MCDM, with a focus on Grey AHP and Grey Entropy methods. This is followed by a detailed explanation of the methodology and data used in the two case studies. Subsequently, the results of each case study are presented and analyzed. The chapter concludes with a discussion of key findings, implications for energy sustainability decision-making, and recommendations for future research.

2. Literature Review

The Grey AHP method integrates grey system theory with the traditional Analytic Hierarchy Process, and the Grey Entropy method combines grey system theory with entropy concepts to objectively determine the weights of criteria in multi-criteria decision-making, particularly useful when dealing with uncertain or incomplete data. This approach has been effectively applied in various energy sustainability contexts. A study by Celik et al. (2023) applied a hybrid Fuzzy AHP and TOPSIS method to convert a LEED-certified education building into a nearly zero-energy building in a cold climate. The methodology helped evaluate energy efficiency scenarios, considering reductions in primary energy use and CO_2 emissions. A novel data fusion model combining Monte Carlo simulations and Grey AHP was developed to select sustainable wastewater treatment systems. Zhian et al. (2024) effectively synthesized expert opinions, ensuring comprehensive decision-making under uncertainty. Roy et al. (2024) applied a grey fuzzy AHP to assess flash flood vulnerability in an ungauged Himalayan watershed, demonstrating the method's versatility in environmental sustainability assessments. A study by Xu and Wang (2024) developed a grey-entropy-catastrophe progression model to assess the resilience of the photovoltaic industry chain. This model provided insights into the industry's adaptability to market uncertainties. Altintas et al. (2020) proposed an extended Grey Relational Analysis method integrated with fuzzy AHP to construct a multidimensional index for ranking the overall energy sustainability performances of OECD countries. Tešić et al. (2023) explained the MARCOS model for the selection of a dump truck using MCDM fuzzy logic. Chatterjee and Chakraborty (2016) described a comparative analysis of various multi-criteria decision-making techniques.

3. Fundamentals of Grey Multiple Criteria Decision Making

3.1 Grey Systems Theory

Grey Systems Theory is a mathematical model which deals with issues involving incorrect, incomplete and ambiguous information when pre-defined decision-making techniques do not work good.

3.2 Multiple Criteria Decision Making (MCDM)

This is a decision making technique, which involves multiple steps - defining criteria and alternatives, weighting of criteria, evaluation and ranking.

3.3 Integration of Grey Systems Theory and MCDM

It is one of the most effective methods compared to the MCDM methods, which require specific data. The combination of Grey Theory with other decision-making techniques creates hybrid models capable of providing decisions for complex problems. A few hybrid models are:

- *Grey-AHP*: Grey values are combined with the Analytic Hierarchy Process to assign weights to each criterion in the presence of incomplete, deficient, or ambiguous data.
- *Grey-Entropy*: Grey values are combined with the Entropy method to assign weights to each criterion in the presence of incomplete, deficient, or ambiguous data.

3.4 Steps in GMCDM

Flow chart shown in Figure 1 represents step by step approach of GMCDM.

a. *Problem Definition*:
   - Clearly define the decision problem, objectives, and constraints.
   - Identify the criteria and alternatives to be evaluated.

b. *Data Collection and Grey Value Assignment*:
   - Collect relevant data for each criterion and alternative.
   - Assign Grey values to represent uncertain or incomplete data.

c. *Weighting of Criteria*: Methods like AHP or Entropy are used to assign weights to each criteria according to their relative importance.

d. *Steps in Grey Association Analysis*:
   - It includes calculation of Grey Association Coefficients (GAC) and Grey Association Grades (GAG) for each alternative considering the assigned weights.
   - The rank is also assigned based on the value of GAG. The alternative with the highest value of GAG will achieve first rank.

e. *Decision Making*:
   - To choose the best suitable alternative and to take decision accordingly, outcomes are analyzed on the basis of rankings.
   - To understand the effects of changing the weights of criteria or ranges of Grey value, sensitivity analysis is done.

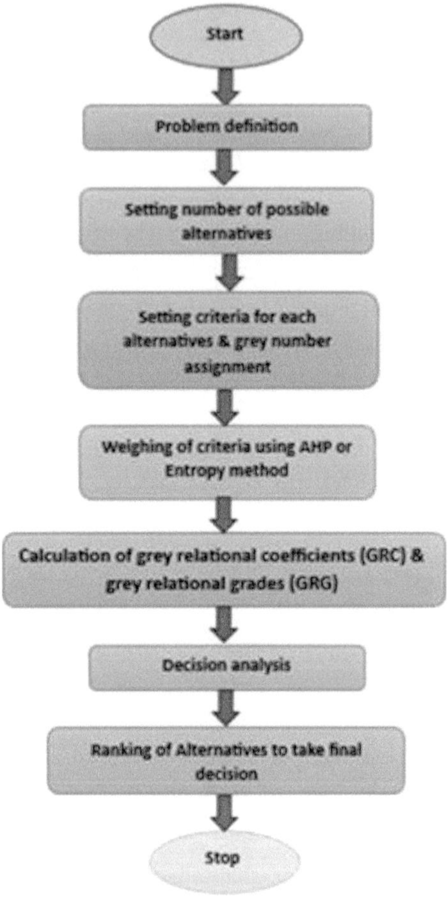

Figure 1. Flow Chart for Grey MCDM.

4. Importance of Grey Values and Grey Relations in Energy System Modeling and Optimization

Since in energy systems appropriate data is often unavailable due to measurement errors, changing climatic conditions, or forecasting ambiguities, Grey values are used to define such parameters as demand, supply, and prices of energy. This study also helps forecast the demand for electricity in the future when limited historical data is available, ensuring better planning and resource allocation. Grey Decision Tables represent implementation cost, efficiency, scalability, and environmental impact of various energy sources using Grey values, showing uncertain outcomes. This thorough study results in a well-informed decision about the most suitable combination of energy sources in the respective situation.

Grey Association Analysis (GAA) evaluates the dependence of energy consumption on factors such as climatic conditions and economic activity. Using this analysis, the production and distribution of energy can be effectively optimized and planned. All the alternatives are ranked using Grey Association Grades, considering the importance of every criterion. Based on this concept, the most suitable option for a renewable energy source or the best location for any energy plant can be determined without losing focus on the goal of energy sustainability—considering all major parameters affecting the importance of an alternative.

5. Formulae Used in GMCDM

5.1 Normalization of Data

Normalize the decision matrix $X = [x_{ij}]$ where i represents criteria and j represents alternatives. For this, VIKOR interval method is used (Nath et al. 2023, Chatterjee and Chakraborty 2016)

Normalization formula for benefit criteria (higher is better)

$$x_{ij}^{*} = \frac{x_{ij} - \min(x_{ij})}{\max(x_{ij}) - \min(x_{ij})} \tag{5.1}$$

Normalization formula for cost criteria (lower is better):

$$x_{ij}^{*} = \frac{\max(x_{ij}) - x_{ij}}{\max(x_{ij}) - \min(x_{ij})} \tag{5.2}$$

5.2 Assign Weights to Each Criterion Based on their Relative Importance

In Grey Multiple Criteria Decision-Making (GMCDM), assigning weights to criteria is crucial as it helps reflect the relative importance of each criterion in the decision-making process. Several methods can be used to determine these weights, such as the Analytic Hierarchy Process (AHP), Entropy Method, or directly using expert judgment.

5.2.1 Steps for Assigning Weights Using AHP

Step 1: Define the Problem and Structure the Hierarchy

- Identify the goal of the decision-making problem.
- List the criteria that will be used to evaluate the alternatives.
- Structure the hierarchy with the goal at the top, followed by the criteria, and then the alternatives at the bottom.

Step 2: Construct the Pairwise Comparison Matrix

In AHP, decision-makers compare each pair of criteria to determine their relative importance. The pairwise comparison matrix A is constructed as follows:

$$A = \begin{bmatrix} 1 & a_{12} & \dots & a_{1n} \\ 1/a_{12} & 1 & \dots & a_{2n} \\ \vdots & \vdots & \vdots & \vdots \\ 1/a_{1n} & 1/a_{2n} & \dots & 1 \end{bmatrix}$$

Here, each a_{ij} of the matrix represents the relative importance of the criterion i to criterion j. Values are assigned based on a scale from 1 to 9 (1 = equally important, 9 = extremely more important).

Step 3: Normalize the Pairwise Comparison Matrix:

Normalize the pairwise comparison matrix by dividing each element by the sum of its column.

Step 4: Calculate the Priority Vector (Weights)

Calculate the priority vector (weights) by averaging the normalized values across each row.

Step 5: Calculate the Consistency Index (CI)

To ensure consistency in the pairwise comparisons, calculate the consistency index (CI). First, calculate λ_{max}

$$\lambda_{\max} = \sum_{i=1}^{n} \frac{(Aw)_i}{w_i} \tag{5.3}$$

Then, calculate the consistency index

$$CI = \frac{\lambda_{\max} - n}{n-1} \tag{5.4}$$

Step 6: Calculate the Consistency Ratio (CR)

Finally, calculate the consistency ratio (CR) to determine the consistency of the pairwise comparisons.

$$CR = \frac{CI}{RI} \tag{5.5}$$

Where RI (Random Consistency Index) is the average CI for a randomly generated pairwise comparison matrix, and its value depends on the number of criteria n.

5.2.2 Steps for Assigning Weights Using Entropy Method

The entropy method is another approach to objectively determine the weights of criteria in decision-making problems. It relies on the concept of information entropy to measure the uncertainty or randomness associated with each criterion.

Step 1: Construct the Decision Matrix

Create a decision matrix X with m alternatives and n criteria:

$$X = \begin{bmatrix} x_{11} & x_{12} & \cdots & x_{1n} \\ x_{21} & x_{22} & \cdots & x_{2n} \\ \vdots & \vdots & \vdots & \vdots \\ x_{m1} & x_{m2} & \cdots & x_{mn} \end{bmatrix}$$

Where x_{ij} represents the value of the ith alternative under the jth criterion.

Step 2: Normalize the Decision Matrix

Normalize the decision matrix X to obtain the normalized matrix $R = [r_{ij}]$

Normalization formula for benefit criteria (higher is better)

$$r_{ij} = \frac{x_{ij}}{\max_j x_{ij}} \tag{5.6}$$

Normalization formula for cost criteria (lower is better):

$$r_{ij} = \frac{\min_j x_{ij}}{x_{ij}} \tag{5.7}$$

Step 3: Calculate the Entropy of Each Criterion

Calculate the entropy E_j for each criterion j (Zhao et al. 2020)

$$E_j = -k\sum_{i=1}^{m} e_{ij} \ln e_{ij} \tag{5.8}$$

Where, $e_{ij} = \frac{r_{ij}}{\sum_{i=1}^{m} r_{ij}}$

Where $k = \frac{1}{\ln m}$ is a constant that ensures $0 \leq E_j \leq 1$.

Step 4: Determine the Degree of Diversification (Redundancy)

Calculate the degree of diversification d_j for each criterion j

$$d_j = 1 - E_j \tag{5.9}$$

Step 5: Calculate the Weights of Each Criterion

Determine the weights w_j for each criterion j (Liang et al. 2022)

$$w_j = \frac{d_j}{\sum_{j=1}^{n} d_j} \tag{5.10}$$

5.3 Calculation of Grey Association Coefficient

$$g_{ij} = \frac{\Delta_{\min} + \rho\,\Delta_{\max}}{\Delta_{ij} + \rho\,\Delta_{\max}} \tag{5.11}$$

Where, Δ_{ij} represents difference between the ideal and actual performance for criterion j and alternative i, Δ_{min} and Δ_{max} are the minimum and maximum differences, and ρ is the distinguishing coefficient (usually $\rho = 0.5$) (Huang et al. 2022).

5.4 Calculation of Grey Association Grades

Aggregate the grey association coefficients (g_{ij}) to obtain the Grey Association Grade

$$G_j = \frac{1}{n}\sum_{i=1}^{n} w_i \cdot g_{ij} \tag{5.12}$$

where n is the number of criteria.

5.5 Rank Alternatives

Rank the alternatives based on the Grey Association Grades. The alternative with the highest grade is considered the best.

5.6 Case Study – 1 (Weighing is Done using AHP Method)

Selection of the best alternative for a sustainable energy project - Select the best alternative for a sustainable energy project considering criteria such as cost, environmental impact, social acceptance, efficiency, and reliability (data is taken hypothetically).

Step 1: Criteria Definition

Identify and define the criteria essential for evaluating alternatives (Table 1):

- *Cost*: Initial investment, operational, and maintenance costs.
- *Environmental Impact*: Carbon footprint, pollution levels, and resource consumption.

Table 1. Data comprising details of five alternatives for sustainable energy.

Alternatives	C1: Energy Production Efficiency	C2: Cost of Implementation	C3: Environmental Impact	C4: Resource Availability	C5: Public Acceptance
Solar (A)	[7, 9]	[5, 7]	[6, 8]	[7, 9]	[8, 10]
Wind (B)	[8, 10]	[6, 8]	[7, 9]	[6, 8]	[7, 9]
Hydropower (C)	[9, 10]	[7, 9]	[5, 7]	[8, 10]	[6, 8]
Geothermal (D)	[8, 9]	[7, 8]	[5, 6]	[7, 8]	[7, 9]

- *Social Acceptance*: Public perception and community engagement.
- *Efficiency*: Energy output relative to input.
- *Reliability*: Consistency and stability of energy supply.

Step 2: Construct AHP Hierarchy

The hierarchical structure for AHP for the selection of Optimal Sustainable Energy Alternatives as per the discussed case is:

Criteria: Cost, Environmental Impact, Social Acceptance, Efficiency, Reliability

Alternatives: Solar Energy, Wind Energy, Hydropower, Geothermal Energy

Step 3: Pairwise Comparisons and AHP Weight Calculation

The relative importance of criteria using a scale from 1 to 9 is finalized by the experts based on surveys. The pairwise comparison matrix for the mentioned criteria is shown in Table 2.

Calculation of Weight Vectors: Compute the normalized weights for criteria using the AHP method (shown in Table 3).

Table 2. The pairwise comparison matrix for the mentioned criteria.

Criteria	C1 (Efficiency)	C2 (Cost)	C3 (Environmental Impact)	C4 (Resource Availability)	C5 (Public Acceptance)
C1 (Efficiency)	1.00	3.00	5.00	7.00	4.00
C2 (Cost)	0.33	1.00	3.00	3.00	2.00
C3 (Environmental Impact)	0.20	0.33	1.00	1.00	1.00
C4 (Resource Availability)	0.14	0.20	0.33	1.00	1.00
C5 (Public Acceptance)	0.25	0.50	1.00	1.00	1.00

Table 3. Weight Table using AHP Method.

Criteria	C1 (Efficiency)	C2 (Cost)	C3 (Environmental Impact)	C4 (Resource Availability)	C5 (Public Acceptance)
Weights	0.516	0.223	0.091	0.067	0.103

Step 4: Grey MCDM Ranking Using GAA

Normalized Data: Convert raw data for alternatives into dimensionless data (Table 4).

Table 4. Normalized data table (Case Study 1).

Energy Source	C1: Energy Production Efficiency	C2: Cost of Implementation	C3: Environmental Impact	C4: Resource Availability	C5: Public Acceptance
Solar (A)	[0, 0.6667]	[0.5, 1]	[0.25, 0.75]	[0.25, 0.75]	[0.5, 1]
Wind (B)	[0.3333,1]	[0.25, 0.75]	[0, 0.5]	[0, 0.5]	[0.25, 0.75]
Hydropower (C)	[0.6667, 1]	[0, 0.5]	[0.5, 1]	[0.5, 1]	[0, 0.5]
Geothermal (D)	[0.3333, 0.6667]	[0.25, 0.5]	[0.75, 1]	[0.25, 5]	[0.25, 0.75]

Grey Association Coefficients (GRC): Calculate the grey association coefficient between each alternative and the ideal solution (Table 5).

Table 5. Grey Association Coefficients Table.

Alternatives	C1: Energy Production Efficiency	C2: Cost of Implementation	C3: Environmental Impact	C4: Resource Availability	C5: Public Acceptance	Grey Association Grade
Solar (A)	0.3098	0.0892	0.0455	0.0334	0.0411	0.5191
Wind (B)	0.2213	0.1115	0.0606	0.0446	0.0514	0.4894
Hydropower (C)	0.1936	0.1487	0.0364	0.0267	0.0685	0.4740
Geo thermal (D)	0.2582	0.1274	0.0331	0.0382	0.0514	0.5083

Calculate Grey Association Grades using GACs and weights.

Step 5: Ranking the alternatives

Following ranking (Table 6) is obtained based on the Grey Association Grades, i.e., GAGs (Ahmed Javed et al. 2022):

Table 6. Final Ranking (Case Study 1).

Energy Source	GAG	Rank
Solar (A)	0.5191	1
Geo thermal (D)	0.5083	2
Wind (B)	0.4894	3
Hydropower (C)	0.4740	4

5.7 *Case Study – 2 (Weighting is Done using the Entropy Method)*

Smart Grid Optimization Using Grey MCDM - Smart grid optimization involves integrating various energy sources, enhancing grid efficiency, reducing costs, and improving reliability and environmental sustainability. This case study illustrates the use of Grey MCDM in optimizing a smart grid represented in Table 7 (data is taken hypothetically).

Table 7. Data comprising details for Smart Grid Optimization.

Alternatives	Cost (C1)	Reliability (C2)	Energy Efficiency (C3)	Environmental Impact (C4)	Scalability (C5)
Traditional Grid Upgrades (A1)	[7, 9]	[8, 10]	[6, 8]	[6, 8]	[7, 9]
Distributed Generation (A2)	[6, 8]	[7, 9]	[7, 9]	[8, 10]	[8, 10]
Energy Storage Systems (A3)	[7, 9]	[8, 10]	[7, 9]	[7, 9]	[6, 8]
Demand Response Programs (A4)	[6, 8]	[7, 9]	[6, 8]	[6, 8]	[8, 10]
Renewable Energy Integration (A5)	[6, 8]	[7, 9]	[8, 10]	[7, 9]	[8, 10]

Step 1: Choosing Criteria for Evaluation:

Criteria: Cost (C1), Reliability (C2), Energy Efficiency (C3), Environmental Impact (C4), Scalability (C5)

Alternatives: Traditional Grid Upgrades (A1), Distributed Generation (A2), Energy Storage Systems (A3), Demand Response Programs (A4), Renewable Energy Integration (A5)

Step 2: Weight calculation using the Entropy method

Table 8 represents Entropy matrix for each alternative under considered criteria.

Table 8. Entropy matrix.

Alternatives	Cost (C1)	Reliability (C2)	Energy Efficiency (C3)	Environmental Impact (C4)	Scalability (C5)
Traditional Grid Upgrades (A1)	0.52304056	0.41671	0.57156	0.46341	0.5004
Distributed Generation (A2)	–0.46341356	–0.5004	–0.5004	–0.54987	–0.41671
Energy Storage Systems (A3)	–0.52304056	–0.41671	–0.5004	–0.52304	–0.57156
Demand Response Programs (A4)	–0.46341356	–0.5004	–0.57156	–0.46341	–0.41671
Renewable Energy Integration (A5)	–0.46341356	–0.5004	–0.41671	–0.52304	–0.41671

Table 9 represents weights calculated using Entopy method.

Table 9. Weight using the Entropy method.

Criteria	Cost (C1)	Reliability (C2)	Energy Efficiency (C3)	Environmental Impact (C4)	Scalability (C5)
WEIGHTS	0.125782065	0.094316266	0.254978344	0.263905548	0.261017778

Step 3: Normalize the Data

Table 10 represents the normalized data for each alternative.

Table 10. Normalized data table (Case Study 2).

Alternatives	Cost (C1)	Reliability (C2)	Energy Efficiency (C3)	Environmental Impact (C4)	Scalability (C5)
Traditional Grid Upgrades (A1)	[0, 0.6667]	[0.3333, 1]	[0, 0.5]	[0.5, 1]	[0.25, 0.75]
Distributed Generation (A2)	[0.3333, 1]	[0, 0.6667]	[0.25, 0.75]	[0, 0.5]	[0.5, 1]
Energy Storage Systems (A3)	[0, 0.6667]	[0.3333, 1]	[0.25, 0.75]	[0.25, 0.75]	[0, 0.5]
Demand Response Programs (A4)	[0.3333, 1]	[0, 0.6667]	[0, 0.5]	[0.5, 1]	[0.5, 1]
Renewable Energy Integration (A5)	[0.3333, 1]	[0, 0.6667]	[0.5, 1]	[0.25, 0.75]	[0.5, 1]

Step 4: Calculate Grey Association Coefficients (GRC)

Using the normalized data, calculate the Grey Association Coefficient (GAC) for each criterion.

Now, Calculate Grey Association Grades (GAG) using GAC and weights as represented in Table 11.

Table 11. Grey Association Coefficients (GAC) – Case Study 2.

Alternatives	Cost (C1)	Reliability (C2)	Energy Efficiency (C3)	Environmental Impact (C4)	Scalability (C5)	GAG
Traditional Grid Upgrades (A1)	0.0755	0.0404	0.17	0.1056	0.1305	0.5219
Distributed Generation (A2)	0.0539	0.0566	0.1275	0.1759	0.1044	0.5183
Energy Storage Systems (A3)	0.0755	0.0404	0.1275	0.132	0.174	0.5493
Demand Response Programs (A4)	0.0539	0.0566	0.17	0.1056	0.1044	0.4905
Renewable Energy Integration (A5)	0.0539	0.0566	0.102	0.132	0.1044	0.4488

Step 5: Ranking the alternatives

Following ranking is obtained based on the Grey Association Grades as shown in Table 12:

Table 12. Final Ranking – Case Study 2.

Energy Source	GRG	Rank
Energy Storage Systems (A3)	0.5493	1
Traditional Grid Upgrades (A1)	0.5219	2
Distributed Generation (A2)	0.5183	3
Demand Response Programs (A4)	0.4905	4
Renewable Energy Integration (A5)	0.4488	5

6. Results and Discussion

In this chapter, we have studied two case studies using Grey AHP method and Grey Entropy method, respectively, and the following results are obtained:

In Case Study 1:

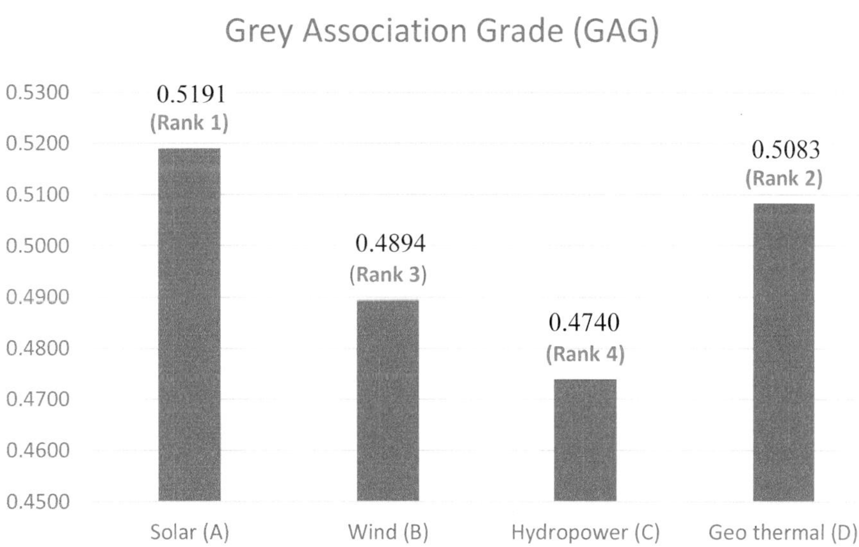

Figure 2. Ranking of alternatives using GAG in Case Study 1.

Ranking of alternatives based on their Grey Association Grades is clearly visible in Figure 2. Higher grades indicate better performance relative to the criteria. In this way, by assessing all the criteria for each alternative to select the best alternative for sustainable energy, it is observed that solar energy is the best option, while hydropower is an avoidable option due to the limited availability of resources. This guarantees the deployment of renewable energy with the greatest efficiency and dependability.

In Case Study 2:

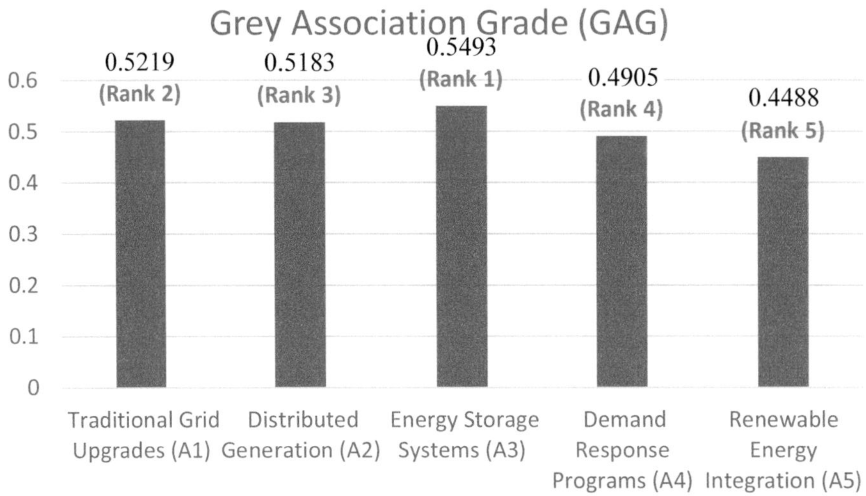

Figure 3. Ranking of alternatives using GAG in Case Study 2.

The alternative with the highest GAG is the optimal choice. Thus, GMCDM can help with the design and implementation of smart grid technologies, including distributed generation, demand response programs, and microgrids. By evaluating a variety of criteria, it helps to maximize grid resilience and performance. It is observed here that with the help of Figure 3, under the given circumstances, the use of energy storage systems is the best alternative. However, as it is clear from the GAG scores that there is not a major difference between the scores of multiple alternatives, a mix of these alternatives can also be implemented in the given case.

7. Conclusion

By managing ambiguities and assessing several criteria, Grey Multi-Criteria Decision Making (GMCDM) is one of the best tools for the optimization of sustainable energy systems. It provides decision-makers with a structured and systematic way to reconcile environmental, social, and economic aspects, improving the selection of choices. Because of its adaptability, GMCDM can be used in a variety of energy-related situations and scales, including smart grid optimization and policy-making for renewable energy integration. Going forward, GMCDM's potential lies in its capacity to adapt to innovation, ease the combination of advanced technologies, and aid in the shift to a more sustainable energy environment. Integrating these computational methods, creating real-time decision support systems, encouraging collaboration platforms, increasing sustainability measures, and conducting international case studies are some future study avenues.

7.1 Future Scope

Grey Multi-Criteria Decision Making's (GMCDM) future depends on its capacity to evolve with new technologies and adapt to intricate cross-disciplinary decision-making. In a rapidly changing world, its capacity to manage uncertainty makes it crucial for making sustainable, well-informed decisions. GMCDM could be applied in next-generation decision-making frameworks such as machine learning-based optimization, neural networks, and data-intensive issues. It could be applied to renewable energy systems to optimize novel technologies, including smart grids, advanced battery storage, and hydrogen fuel cells under uncertain conditions. It works as a good tool in the planning of sustainable urban infrastructures by considering multiple conflicting criteria like energy efficiency, economic feasibility and environmental impact in the context of smart cities. This method can also be adapted by dynamic decision-making processes with evolving data, making decisions more responsive to real-time updates. With the combination of fuzzy logic and Grey Association Analysis (GAA), higher levels of uncertainty and ambiguity could be handled very effectively.

Acknowledgment

I would like to thank the Director, ABES Engineering College, Ghaziabad, for his guidance and motivation to carry out this work.

References

Adali, E. A., Öztaş, G. Z., Öztaş, T. and Tuş, A. 2022. Assessment of European cities from a smartness perspective: An integrated grey MCDM approach. Sustainable Cities and Society 84: 104021.

Ahmed Javed, Angappa Gunasekaran and Amin Mahmoudi. 2022. DGRA: Multi-sourcing and supplier classification through Dynamic Grey Relational Analysis method. Computers and Industrial Engineering 173: 108674.

Altintas, K., Vayvay, O., Apak, S. and Cobanoglu, E. 2020. An extended GRA method integrated with fuzzy AHP to construct a multidimensional index for ranking overall energy sustainability performances. Sustainability 12(4). 1602.

Canbulut, G., Köse, E. and Arik, O. A. 2022. Public transportation vehicle selection by the grey relational analysis method. Public Transport: 1–18.

Celik, A., Sahin, B., Manay, E. and Balin, A. 2023. A study using the hybrid fuzzy AHP&TOPSIS method in the conversion of a LEED-Certified education building into a nearly zero-energy building in a cold climate. Strojniški vestnik-Journal of Mechanical Engineering 69(5-6): 208–223.

Chatterjee, P. and Chakraborty, S. 2016. A comparative analysis of VIKOR method and its variants. Decision Science Letters 5(4): 469–486.

Huang, C. Y., Lin, Y. C., Lu, Y. C. and Chen, C. I. 2022. Application of grey relational analysis to predict dementia tendency by cognitive function, sleep disturbances, and health conditions of diabetic patients. Brain Sciences 12(12): 1642.

Javed, S. A., Gunasekaran, A. and Mahmoudi, A. 2022. DGRA: Multi-sourcing and supplier classification through Dynamic Grey Relational Analysis method. Computers & Industrial Engineering 173: 108674.

Liang, M., Li, W., Ji, J., Zhou, Z., Zhao, Y., Zhao, H. et al. 2022. Evaluating the comprehensive performance of 5G base station: A hybrid MCDM model based on Bayesian Best-Worst method and DQ-GRA technique. Mathematical Problems in Engineering 2022(1): 4038369.

Nath, B., Basumatary, B., Wary, N., Basumatary, U. R., Basumatary, J., Rokhum, S. L. et al. 2023. Agricultural waste-based heterogeneous catalyst for the production of biodiesel: a ranking study via the VIKOR method. International Journal of Energy Research 2023(1): 7208754.

Roy, D., Dhar, A. and Desai, V. R. 2024. A grey fuzzy analytic hierarchy process-based flash flood vulnerability assessment in an ungauged Himalayan watershed. Environment, Development and Sustainability 26(7): 18181–18206.

Tešić, D., Božanić, D., Puška, A., Milić, A. and Marinković, D. 2023. Development of the MCDM fuzzy LMAW-grey MARCOS model for selection of a dump truck. Reports in Mechanical Engineering 4(1): 1–17.

Xu, L. and Wang, Y. 2024. Resilience evaluation model of photovoltaic industry chain based on grey-entropy-catastrophe progression method: a case study of Jiangsu province. Grey Systems: Theory and Application 14(3): 491–506.

Zhao, S., Gong, M., Liu, T., Fu, H. and Tao, D. 2020. Domain generalization via entropy regularization. Advances in Neural Information Processing Systems 33: 16096–16107.

Zhian, T., Hashemi Monfared, S. A., Rashki, M. and Azizyan, G. 2024. Enhancing decision fusion for wastewater treatment system selection using Monte Carlo simulation and gray analytic hierarchy process. Water 16(12): 1709.

Chapter 6

Clean Energy Transitions Towards Achieving Net Zero Emissions Targets in Aviation Industry

A Review of Issues and Pathways

Mukesh Pandey,[1,*] *Divya Shukla*[2] and *Huw Davies*[3]

1. Introduction

The aviation industry's pursuit of sustainability has become increasingly urgent in light of growing pressure to achieve net-zero emissions. While this challenge is now widely acknowledged, the underlying reasons for its urgency and the means through which it may be addressed are still not sufficiently developed in the literature. To situate the discussion meaningfully, it is first necessary to understand the broader climate context within which aviation decarbonisation is being debated.

The rapid accumulation of greenhouse gas (GHG) emissions has driven a marked rise in global temperatures, with the average global surface temperature during 2011–2020 reaching 1.1°C above the pre-industrial level (IPCC 2023). Continued emissions are expected to intensify this trend further. Current projections indicate that global warming is likely to reach 1.5°C in the near term and could rise to 2.7°C by the end of the century if present trajectories persist (IPCC 2023).

[1] Assistant Professor in Aviation Management, Coventry University UK.
[2] Senior Lecturer in Leadership and Organisational Behaviour University of Northampton UK.
[3] Assistant Professor at Centre for Future Transport and Cities, Coventry University UK.
* Corresponding author: ad2609@coventry.ac.uk

DOI: 10.1201/9781003475385-6

As global warming continues to intensify, the risks it poses to human societies and natural ecosystems are becoming more serious, more widespread and increasingly interconnected. In response to these growing threats, 196 countries adopted the Paris Agreement, a legally binding international treaty, at the UN's Climate Change Conference (COP21) on 12th December 2015 held in Paris. The agreement aimed to keep the rise in global average temperature below 2°C from pre-industrial level, while also making efforts to limit it to 1.5°C (Paris Agreement 2015). Achieving this ambition will require a sharp reduction in global greenhouse gas emissions, with cuts of around 45% by 2030 and the achievement of net zero by 2050 (IPCC 2023).

Net zero refers to a state in which GHG emissions are reduced to a level that can be balanced through natural or technological absorption. Reaching this point demands careful attention to the difficulties faced by different sectors so that credible pathways for sustainability transitions can be identified. Within this broader debate, the transport sector requires scrutiny. In 2019, it was the fourth-largest source of greenhouse gas emissions globally, following the power, industrial and agricultural sectors (IEA 2020). The sector was responsible for almost 15% of total global greenhouse gas emissions and about 23% of energy-related carbon dioxide emissions, with direct emissions reaching approximately 8.9 Gt CO_2-eq in that year (IEA 2020). Within overall transport emissions, road transport accounted for 69%, while aviation and shipping each contributed 11%, rail represented 1%, and the remaining share came from other modes together with indirect emissions (IPCC 2020).

Although emissions from transport fell sharply during the COVID-19 pandemic, this decline proved temporary. By 2022, transport emissions had largely returned to their pre-pandemic level, reaching around 8 Gt CO_2-eq (IEA 2023). Across the longer period from 1990 to 2022, transport emissions increased at an average annual rate of 1.7%. To align the sector with net-zero ambitions, however, carbon dioxide emissions must decline by more than 3% annually by 2030 (IEA 2023). This highlights the urgency of implementing substantial mitigation measures across all transport sub-sectors.

Within this broader transport picture, aviation contributes a comparatively smaller share of emissions, yet it remains one of the most difficult sectors to decarbonise. This difficulty arises from several interrelated factors, including the limited maturity of clean aviation technologies, continued growth in air travel demand, and the strict safety, certification and regulatory requirements that shape the industry. Aviation is currently responsible for approximately 2.5% of global carbon dioxide emissions and is estimated to contribute around 4% of total greenhouse gas emissions (Lee et al. 2021). Moreover, current projections suggest that aviation emissions may continue to rise, as growth in revenue passenger kilometres is expected to outpace the rate of decarbonisation (Gössling and Humpe 2024). Against this backdrop, there is a clear need to examine the present emissions profile of the aviation sector and to assess the techno-economic prospects of the clean technologies proposed to support its transition.

A growing body of work suggests that substantial reductions in aviation emissions will depend on transformative advances in aircraft technology (Dray and Schäfer

2022). However, the practical viability of such solutions remains contested. Much of the literature continues to point to persistent political, societal, technological and economic barriers that may slow or constrain implementation (Åkerman et al. 2021, Bergero et al. 2023, Gössling and Humpe 2024). Therefore, the transition cannot be understood solely as a matter of technological substitution. Meaningful progress depends on wider systemic change involving policy, market structures, regulation, infrastructure, industrial actors, user practices and broader societal acceptance. A socio-technical perspective is therefore especially relevant, as it enables attention to be given to the interdependencies between technological development and the wider social environment in which it unfolds. In the context of aviation, such a perspective is important for understanding how sustainable transition pathways may emerge, stabilise and scale.

The complexity of sustainability transitions further reinforces the value of the multi-level perspective developed by Geels and others. This framework provides a useful means of examining how niche innovations interact with established regimes and broader socio-technical landscapes, and how change occurs through these dynamic relationships (Geels 2002, Geels 2005). Although previous studies on aviation sustainability have adopted systems-oriented thinking (Abram et al. 2022, Agusdinata et al. 2011), relatively limited attention has been paid to the significance of a socio-technical perspective in this area. The chapter therefore aims to recognise the importance of this perspective in understanding and supporting aviation's sustainability transition.

This chapter is based entirely on secondary sources and draws on evidence from research articles and industry reports. The following section examines the current emissions status of the aviation industry and outlines the principal transition pathways for reducing emissions. The third section evaluates the readiness of clean technology options from a techno-economic perspective. The fourth section discusses the role of a socio-technical approach in enabling a faster and more coherent sustainability transition within the aviation industry. The chapter also identifies potential clean energy transition pathways that may support the sector's progress towards net zero emissions, before concluding with the main findings and their implications.

2. Current State of the Emissions in Aviation Industry and Transition to Net Zero Emissions Target

2.1 Current State of Aviation Industry's Emissions

The aviation industry is categorised as a hard-to-abate sector because its operations continue to depend strongly on liquid hydrocarbon jet fuel. The transition to aircraft based on cleaner fuel technologies is likely to be slow, with industry estimates suggesting a development and deployment period of at least 10 to 15 years (ATI 2022). The climate impact of aviation is also not limited to carbon dioxide alone. Although CO_2 from jet fuel combustion accounts for roughly one-third of effective radiative forcing, a substantial share arises from non-CO_2 emissions, including nitrous oxide, water vapour, sulphur dioxide and soot (Lee et al. 2021). These emissions differ in their atmospheric behaviour: residual CO_2 has a long-lasting warming effect that can endure for thousands of years, while non-CO_2 effects are comparatively short-lived and generally fade within a few decades. This makes it clear that achieving net zero in aviation will require mitigation strategies that tackle both CO_2 and non-CO_2 emissions together.

In 2019, aviation produced 1.03 Giga tonnes (Gt) of CO_2, contributing up 3.1% of global CO_2 emissions. When non-CO_2 effects are included, this figure rises to 1.7 Gt of CO_2 equivalent (eq) (IEA 2020). Although emissions from air travel dropped 40% in 2020 due to the COVID-19 pandemic. The rapidly recovering demand of aviation industry will take emissions level to 2019 in the future. The growth of aviation industry will drive the CO_2 emissions to as high as 1.9 Gt CO_2 per year in 2050 or 3.4 Gt CO_2 eq when non-CO_2 forcing is included, approximately twice from the current level (IEA 2023).

As indicated in Figure 1 the aviation industry had a multi-decade growth in CO_2 emission. It is striking to notice that the average rate of CO_2 emission growth for period 1960–2018 was 15 MT CO_2 per year while average rate of CO_2 emission growth for period 2013–2018 was 44 MT CO_2. It can be inferred that the growth rate of aviation emissions in the recent years has increased by 300%. Also, the cumulative emissions of global aviation in 2018 were 32.6 billion tonnes of CO_2, of which approximately 50% were emitted in last two decades. Hence from the above statistics it is evident that global CO_2 emissions from aviation sector is increasing at an increasing rate. Therefore, it is high time to implement the available clean technology option in order to achieve the net zero emissions target.

Figure 1. Global CO_2 emissions from aviation. (Source: Reprinted with permission from Ritchie, H. 2024. https://ourworldindata.org/).

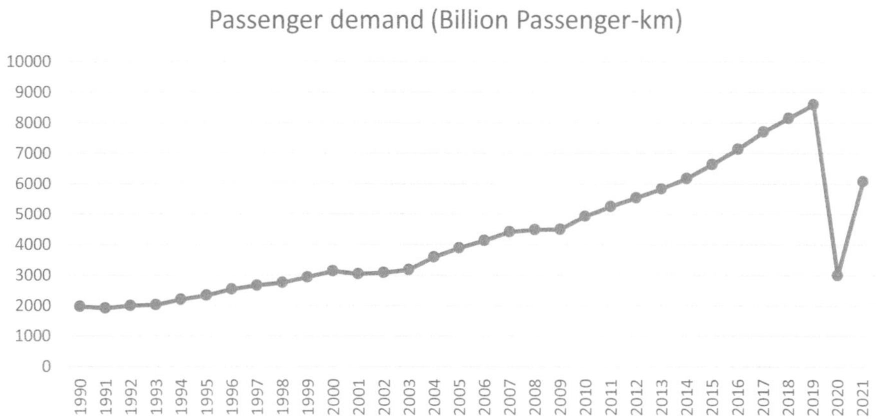

Figure 2. Passenger demand-revenue passenger kilometre in USD. Source: Data from Bergero et al. (2023).

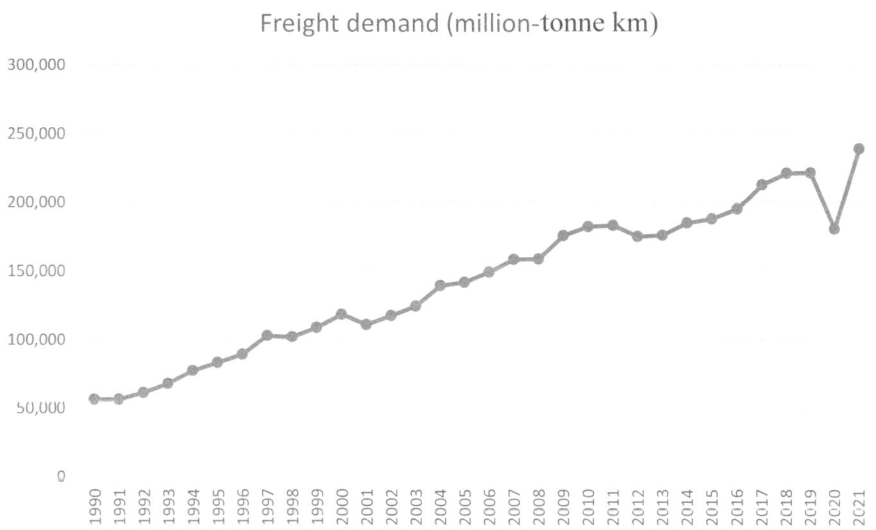

Figure 3. Freight demand-million tonne kilometre. Source: Data from Bergero et al. (2023).

It can be inferred from Figures 2 and 3 that between 1990 and 2019, air travel and cargo transport experienced a fourfold increase in demand. During the same period, aviation's energy efficiency improved twofold. As indicated in Figure 4, in 1990, carbon emissions per passenger-kilometer stood at 357 grams, but by 2019, they had declined to 157 grams (Bergero et al. 2023). These efficiency gains resulted from advancements in aircraft technology and design, larger planes carrying more passengers, and an improved load factor.

While efficiency advancements have helped offset some of the emissions caused by rising demand, overall CO_2 output still grew. Global aviation emissions were

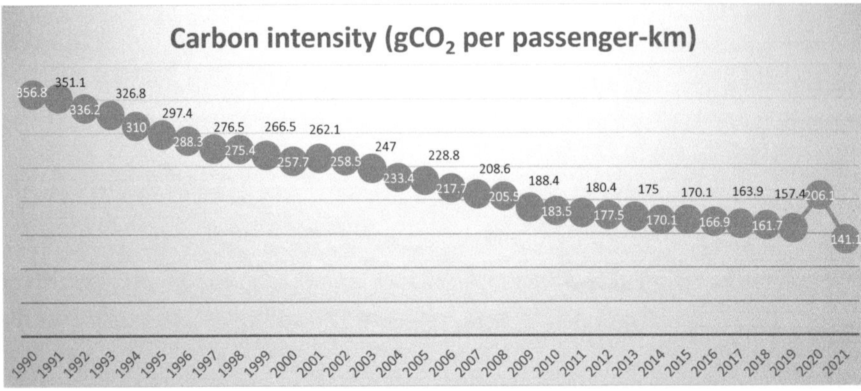

Figure 4. Carbon intensity- CO_2 gm emitted per passenger Km. Source: Data from Bergero et al. (2023).

approximately 0.5 billion tonnes in 1990, reaching around 1 billion tonnes by 2019. The scale of the challenge facing aviation can be seen through the remaining carbon budgets, defined as the amount of greenhouse gases that can still be emitted before critical temperature thresholds are exceeded. According to the IPCC, the remaining carbon budgets across all sectors are about 400 Gt CO_2 for limiting the increase in average global temperature to 1.5°C, 700 Gt CO_2 for 1.7°C, and 1150 Gt CO_2 for 2.0°C (Gössling et al. 2025). If all sectors are expected to reach net zero by 2050 and the remaining budget is shared evenly, aviation, which currently accounts for 2.5% of global CO_2 emissions, would have a carbon allowance of 10 Gt CO_2 under the 1.5°C average global temperature limit and 28.75 Gt CO_2 under the 2.0°C limit (Gössling et al. 2025). As shown in previous statistics, the current rate of annual CO_2 emission in aviation is nearly 1 Gt per year, this is evident that if no mitigations are put in place then after a decade the industry will be operating beyond the net zero carbon budget. Hence, this is a high time to accelerate the emissions reduction measures and to be on transition trajectory of net zero emissions target.

2.2 Regulatory Resolutions and Commitments Towards Achieving Net Zero Emissions Targets

Efforts to curb aviation emissions have increasingly been shaped by international policy and industry commitments. In 1997, ICAO initiated discussions and agreements for addressing emissions from international aviation. After prolonged negotiations among member states, this led to the adoption of the Carbon Offsetting and Reduction Scheme for International Aviation (CORSIA) in 2016 (ICAO 2019). The scheme was designed to stabilise emissions from international aviation at the 2019 level (ICAO 2019). Airlines operating above this threshold are expected to compensate for the excess through approved carbon offsets or through the use of sustainable aviation fuels (SAFs).

Despite this policy development, CORSIA has attracted substantial criticism in the literature. A few scholars argue that the scheme lacks robustness and credibility as a long-term decarbonisation instrument (Gössling and Lyle 2021, Dray and Schäfer 2021, Warnecke et al. 2019, Scheelhaase and Maertens 2020). Concerns centre on its exclusion of emissions at the baseline year, its non-binding legal character, and its failure to account for non-CO_2 climate effects (Dray and Schäfer 2021). Its heavy dependence on offsetting has also been questioned, particularly where forest-based credits are involved, given the growing exposure of such carbon sinks to wildfire risk (Pausas and Keeley 2021). In practice, this reliance on compensation mechanisms may weaken incentives for more fundamental technological and fuel transitions within the aviation sector.

Momentum around net zero gathered further at the 77th IATA Annual General Meeting in 2021, where member airlines committed to achieving net-zero carbon emissions from their operations by 2050 (IATA 2021). According to IATA, meeting this goal would require the abatement of 21.2 gigatonnes of CO_2 between 2021 and 2050 and would demand investment on the order of USD 2 trillion across the industry (IATA 2021). A similar direction was later reinforced at ICAO level, where 184 member states endorsed the long-term aspirational goal (LTAG) of net-zero carbon emissions from international aviation by 2050 (ICAO 2022). However, although LTAG signals collective ambition, it does not impose specific reduction obligations on individual states or airlines (ICCT 2023).

Alongside these international commitments, several governments have introduced national or regional measures to accelerate the transition. In the United States, the Inflation Reduction Act of 2022 provided USD 3.3 billion through tax incentives and competitive grants to expand SAF production (IEA 2023). This intervention supports the national objective of producing 3 billion gallons of SAF by 2030 and the longer-term aim of replacing fossil-derived jet fuel with SAF by 2050 (IEA 2023, Braun et al. 2024). Within Europe, the European Union agreed the ReFuelEU Aviation framework in 2023 under the Fit for 55 package (IEA 2023). This measure requires a 2 per cent SAF blend at EU airports from 2025 and sets a trajectory towards 70 per cent by 2050 (European Commission 2023). Similarly, the United Kingdom committed GBP 165 million in 2022 to support SAF development following the Jet Zero strategy, with the intention of enabling construction to begin on at least five commercial SAF plants by 2025 (IEA 2023, DfT 2023).

Taken together, these developments show that aviation governance is moving more clearly towards emissions reduction, although, the progress remains uneven. Much of the current policy landscape still depends on voluntary commitments, with ReFuelEU Aviation standing out as one of the few binding instruments. Regulatory action has also been concentrated largely in a small number of developed economies, while broader global alignment remains limited. A wider policy mix may therefore be needed to speed up the adoption of cleaner energy pathways. This could include support for SAF, hydrogen and renewable energy production, alongside fiscal measures, blending mandates, and restrictions on fossil fuel use. At present, however, a coherent and coordinated transition strategy across the main system actors is still not fully evident. The next section therefore examines the main clean energy pathways that could support aviation's transition towards net-zero emissions.

2.3 Transition Pathways Towards Achieving Net-Zero Emissions Targets in Aviation

Most reports suggest that net zero is unlikely to be delivered through a single solution and will instead depend on a combination of actions, including wider use of SAF blends, advances in existing aircraft technology, better operational efficiency at airports and in air traffic management, the introduction of battery- or hydrogen-powered aircraft, and carbon offsetting through market-based measures (Braun et al. 2024, IATA 2021, ATAG 2021). The Figure 5 below presents the contributions of technology option which IATA predicts, will help in reducing CO_2 emissions to realise net zero emission targets.

Among all these options, SAF has been identified as most viable option which could be rolled out on large scale in short term. Waypoint 2050 report has predicted that SAF would contribute upto 71% in CO_2 emission reduction by 2050 (ATAG 2021) while IATA has identified that SAF would help in abating 65% of CO_2 emissions by 2050 (IATA 2021).

It is now well established that SAF can be blended with conventional jet fuel, and this has been demonstrated through a range of flight tests. Recently 100% SAF powered flight was successfully conducted by Virgin Atlantic, popularly known as Flight 100, which travelled London Heathrow airport to JFK airport New York in 2023 (Virgin Atlantic 2023). However, SAF use in regular flight operations remains very limited, and at present it is used on less than 1% of flights across Europe (Braun et al. 2024, EASA 2024). The main challenges for the low uptake of SAF are weak supply of SAF, infant supply chain of SAF and high production cost which has

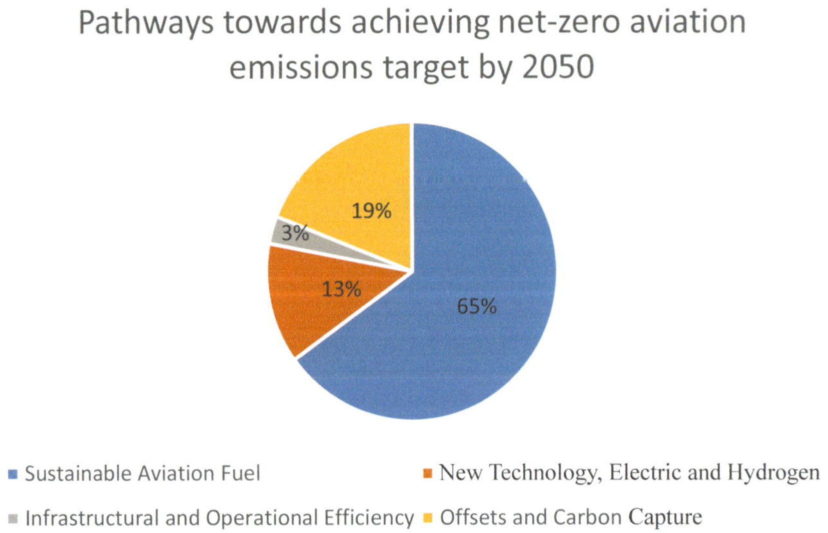

Figure 5. Pathways towards achieving net zero emission targets by 2050. Source: (Data from IATA 2021).

resulted in substantially higher market price for SAF in comparison to the incumbent fuel. In the subsequent section, each available options of emission reduction will be explored.

3. Alternative Clean Energy Technologies in Aviation and Their Techno-Economic Considerations

The clean technology options in aviation includes the SAF, Synthetic fuels, Hydrogen fuels and Battery Electric. This section will evaluate the technology readiness, scalability and techno-economic analysis of these technologies.

3.1 Sustainable Aviation Fuel (SAF)

SAF are bio-material derived Synthesised Paraffinic Kerosene (SPK) which could be blended to fossil fuel-based jet fuel and utilised on existing aircrafts without any modification to the jet engines. Currently, SAF is the only technology options which is ready to be deployed in aviation sector. Hence all the reports of IATA, ATAG, Eurocontrol have relied heavily on the use of SAF towards achieving net zero emission targets (ATAG 2021, IATA 2021, Eurocontrol 2022). Watson et al. (2024) suggest there are seven SAF production pathways which has been certified by American Society for Testing and Materials (ASTM) which includes:

a. Fischer-Tropsch (FT-SPK),
b. Fischer-Tropsch synthetic kerosene with aromatics (FT-SPK/A),
c. Hydroprocessed esters and fatty acids (HEFA-SPK),
d. Hydrocarbon hydroprocessed esters and fatty acids (HC-HEFA-SPK)
e. Synthetic iso-paraffins from fermented hydroprocessed sugar (HFS–SIP),
f. Alcohol-to-jet (ATJ-SPK) and
g. Catalytic hydrothermolysis (CHJ-SPK)

Based on the similarity of production processes, the seven production pathways of SAF could be clubbed into five broad categories as indicated in Figures 6 and 7.

The FT-SPK pathway utilises the feedstock of forestry residues, grasses or municipal solid waste which undergoes gasification and Fisher Tropsch (FT) conversion to Synthetic Paraffin Kerosene (SPK). The Fischer–Tropsch process has been commercially used to convert coal and natural gas into syngas, although its application to biomass is more challenging because of the variable composition of the feedstock (Watson et al. 2024). Consequently, biomass must undergo pre-treatment, which can be both necessary and costly, to enhance syngas quality by increasing carbon and hydrogen constituents while eliminating contaminants (Watson et al. 2024, Goh et al. 2022). The Technology Readiness Level (TRL) indicates the stage of technological development leading to commercialization, while the Fuel Readiness Level (FRL) quantifies the risks of the creation of new fuels. The TRL and FRL of this pathway is 6–8 and 6–7 respectively, indicating this pathway has still not achieved the full commercialisation level (Watson et al. 2024, Dahal et al. 2021).

HEFA-SPK pathway uses the feedstock of used cooking oil, tallow, camelina, soyabean oil which contains fatty acids and fatty acid esters. The feedstock undergoes the oil extraction, hydro-processing, hydro-cracking and distillation to produce SPK. HEFA is the only SAF technology that has achieved the complete commercial deployment, having a Technology Readiness Level (TRL) and Fuel Readiness Level (FRL) of 9 (Watson et al. 2024, Dahal et al. 2021).

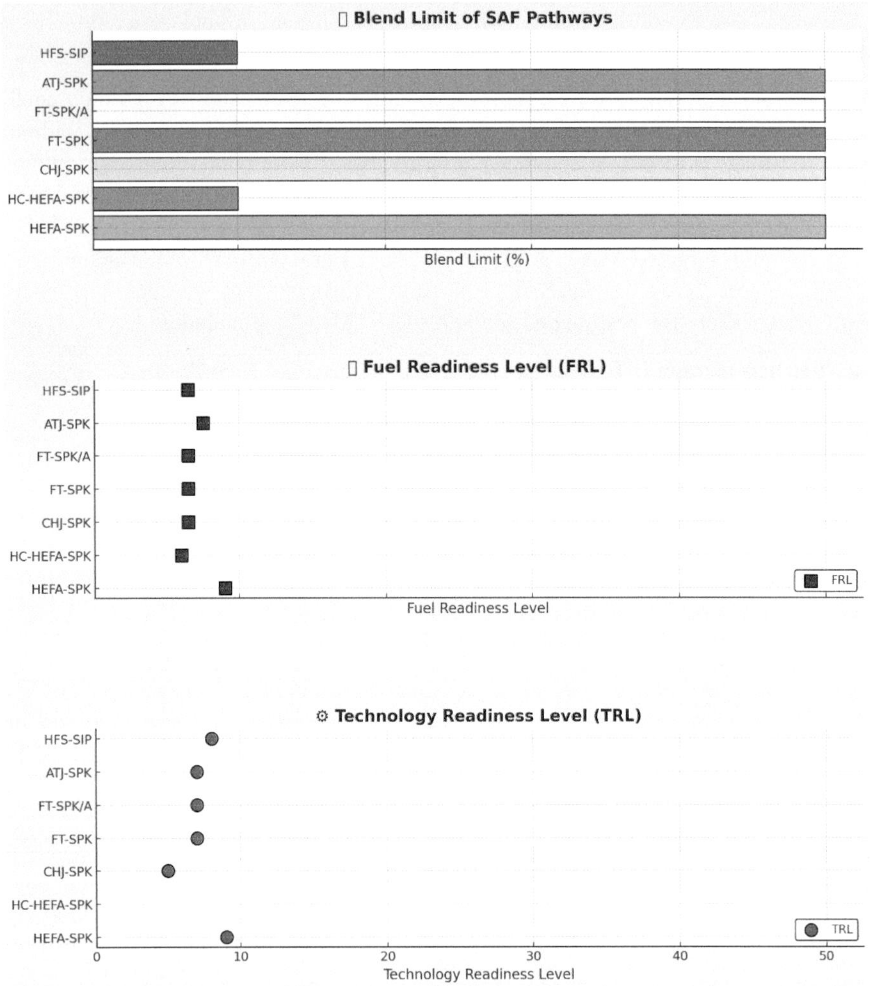

Figure 6. SAF production pathways, TRL, FRL and blend limit. Source: (Data from Watson et al. 2024, Abrantes et al. 2021, Dahal et al. 2021).

HFS-SIP pathway relies on feedstock like cane, sweet, sorghum, sugar beets, tubers, field corn which are pre-treated, undergoes enzymatic hydrolysis, followed by anaerobic fermentation, hydro-processing, hydro-cracking and distillation to produce the bio-jet fuel. The HFS–SIP process is the only method that generates SIP fuel. Its Fuel Readiness Level (FRL) falls between 5 and 8, mainly because SIP and Jet A have different fuel properties, with viscosity being a key difference (Watson et al. 2024, Dahal et al. 2021).

CHJ-SPK pathway depends on Triglyceride-based feedstocks such as plant oils, waste oils, algal oils, soybean oil, jatropha oil, camelina oil, carinata oil and tung oil which undergoes hydrothermal conversion to produce SAF. Compared with HEFA, the CHJ-SPK route requires less hydrogen, but its technological development remains less advanced, with a TRL of 4–6, largely because it depends on demanding operating conditions such as temperatures of 450–475°C and pressures of 210 bar (Watson et al. 2024, Dahal et al. 2021, Eswaran et al. 2021).

ATJ-SPK needs agricultural residue such as stover, grasses, forestry slash, crop straws, corn grains and herbaceous energy crops which undergoes thermo-chemical conversion of alcohols to derive the SPK. The ATJ process shows significant promise for commercial expansion, particularly with bioethanol as a feedstock, due to its global availability, cost-effectiveness, and market dominance compared to other alcohols (Goh et al. 2022). However, since bioethanol is widely utilized in the petrochemical industry as a required additive in automotive fuel, scaling up ATJ production could lead to competition between the aviation and ground transportation sectors for feedstock resources (Doliente et al. 2020).

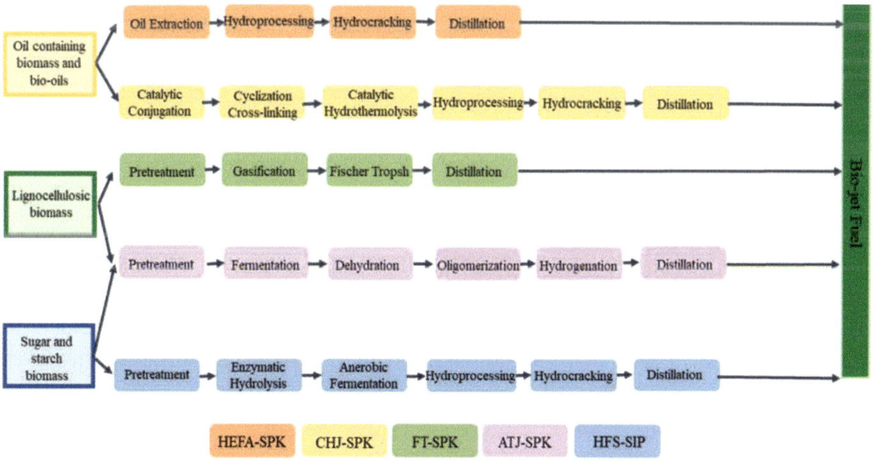

Figure 7. Main process steps for five major SAF production pathways. Source: (Data adopted from Watson et al. 2024).

3.2 *Evaluation of Economic and Environmental Performance of SAF Pathways*

In this section the techno-economic analysis of SAF production pathways is presented. Watson et al. (2024) has conducted the review of 58 research papers, who have done the techno-economic analysis of various SAF pathways. The outcome of this research very well compares and summarises the intricate interplay of economics and environmental considerations for each SAF production pathways. Table 1 in the appendix adopted from Watson et al. (2024) summarises the average economic and environmental performance of each SAF pathway by comparing average minimum jet fuel selling price (MJSP) per litre and average GHG emissions (Gt CO_2 eq per megajoule) against each production pathway and conventional jet fuel. In 2021, the average price of convention jet fuel was USD 0.5 per litre and GHG emission was 89 g CO_2 per MJ (IEA 2021, Prussi et al. 2021).

Economically, HEFA-SPK pathway are the most economical technology still 120% more expensive than conventional jet fuel and are reducing 27% decrease in the GHG emissions. The second most economical SAF production pathway is CHJ-SPK which is 160% more expensive than conventional jet fuel resulting in 77% decrease in the GHG emissions. HEFA and CHJ pathways are economical because of established bio-diesel supply chain. FT-SPK pathway offers maximum reduction of the emissions of 92% however, it is 320% expensive than the jet fuel and relatively low TRL.

The above analyses are based on averages and are scarce to compare the SAF technologies as other factors such as feedstock variations, geographical variations and methodologies to estimate the cost could have influence on the results. For example- In US, SAF produced adopting ATJ pathways using the corn stover feedstock costs USD 1.25 per litre while SAF produced using the corn ethanol feedstock cost USD 0.53 per litre (Park et al. 2022). Overall, the most appropriate SAF technology will vary by country, as its viability depends on factors such as feedstock cost and availability, co-products, and the availability of land and water resources (Watson et al. 2024).

3.3 *Hydrogen Fuel*

The mitigation of non-CO_2 emission, which causes two-third of effective radiative forcing in warming is also essential in order to achieve the net zero in aviation. Although the use of SAF helps significantly in abating the CO_2 emissions, however reduction of non-CO_2 emissions remains still limited with use of SAF. A battery powered or hydrogen fuelled aircraft are the only technology options at the moment which has zero emissions. In this section, the current state of development of hydrogen powered aircraft and its techno-economic considerations will be presented.

The use of hydrogen as a fuel in aircraft requires a massive design and technical challenges due to its higher specific energy and unique storage challenge due to its lower energy density. The property of hydrogen with the conventional jet fuel is compared in Table 2 in appendix. The highest energy density obtained is in liquid hydrogen but it is four times lower than Jet fuel (Alder and Martins 2023). Hence the use of hydrogen will need a specialised higher volume tank in relation to conventional Jet fuel and will incur a weight and drag penalty (Alder et al. 2023).

Several design concepts and feasibility studies on the use of hydrogen as fuel on aircraft have been carried out and are summarised in the Table 2. Depending on the propulsion system considered, feasibility studies have focused either on gaseous hydrogen, used in fuel cells to produce electricity for the electric powertrain, or on liquid hydrogen, used directly in gas turbines to generate thrust (Tiwari et al. 2024). As liquefied hydrogen is four times less dense than Jet A-1 hence it requires approximately four times higher volume of fuel storage. The increased volume storage for hydrogen requires greater research for integration of aero-propulsive and aero-structural systems of the aircraft.

Hydrogen has attracted interest as an aviation fuel for many decades. Work in this area dates to 1937, and several military programmes in the 1950s and 1960s explored its potential, including a 1957 flight test using a modified B-57 bomber (Tiwari et al. 2024). Interest grew further after the oil crises of the 1970s. Lockheed Martin's TU-155 project first developed short-range aircraft designs and later completed a flight test in 1988 (Tiwari et al. 2024). The European Commission also launched the Cryoplane project, which examined the adaptation of the A310 for hydrogen use (Tiwari et al. 2024). In later years, NASA demonstrated hydrogen application in its X-43 experimental hypersonic aircraft in 2004, while Boeing explored the technology through the Phantom Eye project between 2010 and 2014. More recently, Embraer has proposed sub-regional hydrogen fuel cell and regional hydrogen turboprop aircraft concepts that may enter service around 2035 to 2040, and Airbus, through its ZEROe project, has presented short- and medium-range hydrogen-powered aircraft concepts expected to enter the market from 2035 onwards (Tiwari et al. 2024, Alders et al. 2023).

Only a few regional hydrogen fuel cell aircraft prototypes have been developed so far, with ZeroAvia and H2Fly likely to begin flying in the next few years (Alders and Martins 2023). For example: Zero Avia is modifying Dornier 228 with hydrogen fuel cell prop aircraft with certifications expected in 2025. While H2Fly is converting Dornier 328 with fuel cell-based propulsion system, to fly a 40 passenger regional aircraft by 2025. These projects appear promising and it is expected that hydrogen fuel cell based small aircraft may enter market by 2030 (ATAG 2021). The midrange hydrogen based aircraft are still on low TRL. The Airbus Zero-e projects are at TRL 3–4 early prototype stage which requires validation (Ravichandran 2024). Due to low TRL level, Airbus predicts that hydrogen aircraft will fully commercialise by 2045 (Ravichandran 2024).

3.3.1 Evaluation of Economic and Environmental Performance of Hydrogen Fuelled Aircrafts

As the technology of Hydrogen powered aircrafts are not matured yet, it's uncertain to predict the specific cost. However, based on the review of existing studies this section will explore the cost scenarios of the technology and compare it with the incumbent. Broadly the cost of hydrogen based aircraft could be broadly segregated into research & development, acquisition, maintenance and fuel cost. Tiwari et al.

(2024) observed that for short to medium range hydrogen aircraft the acquisition expense will be 12–13% higher than the incumbent due to hydrogen storage tank or fuel cell and system integration. Although hydrogen burns more cleanly than the jet fuel, therefore this will result the engine to have 25% more lifetime (Brewer et al. 1978). However, the operational challenge of running hydrogen tank and fuel systems could increase the cost (Alder et al. 2023). The fuel cost of hydrogen is another significant cost. In order to make this alternative viable, there is need to work upon both unit cost of fuel and the fuel efficiency of the engine.

Mukhopadhaya and Rutherford (2022) estimated that hydrogen combustion in turboprop and narrowbody aircraft could increase energy use by about 10% when tank gravimetric efficiency is between 20% and 35%. In contrast, 10% reduction in energy demand for long-range aircraft is estimated when tank gravimetric efficiency is around 75% (Alder et al. 2023). Tiwari et al. (2024) reported estimates within this range and suggested that hydrogen becomes more efficient as flight range increases. The use of fuel cell propulsion could further improve performance, with the potential to double energy efficiency and make hydrogen a more attractive option.

The other aspect of the arriving at a viable fuel cost of hydrogen is to reduce unit cost of the fuel. To illustrate this, it is significant to understand the underlying process. The renewable hydrogen is produced by electrolysis process requiring electrolyser to run by renewable electricity. The electrolysis process is approximately 60%–80% efficient hence production of 1 MJ of hydrogen requires approximately 1.5 MJ of electricity (IEA 2019). After the production, hydrogen requires liquefaction which requires additional renewable electricity. The cost of renewable electricity in whole process is very substantial which is estimated to decrease in due course with the augmented initiatives renewable resources scale up. Many government-funded projects including United States Hydrogen Shot Programme have objectives to reduce the cost of clean hydrogen to USD 1/kg (Tiwari et al. 2024, Alder et al. 2023). Although, many studies have that the cost of hydrogen will be in range of USD 2 to 7 per kg (Zhou and Searle 2022). Hence, to conclude there is a need to enhance the efficiency of hydrogen-based aircraft, at the same time integrated solutions from the several actors needed to find solutions to reduce the per unit cost of renewable energy and hydrogen so that the Hydrogen fuel aircrafts become cost effective to disrupt the incumbent regime.

3.3.2 Evaluation of Environmental Performance of Hydrogen Fuelled Aircrafts

The most advantage of hydrogen aircrafts are they do not emit any CO_2, hence it becomes the competitive alternative to reduce the long-lived warming effect from aviation (Alders and Martins 2023). But hydrogen combustion based aircraft emits NOx which is a short live non carbon polluter. However, a few studies suggest that harmful effects of non-CO_2 polluters could be reduced by modifying the combustion properties.

On the contrary, the Fuel cell based hydrogen aircraft produced no NOx as well (O'hayre et al. 2016). Hence it becomes a very advantageous alternative clean fuel reducing the non-CO_2 emissions. Although hydrogen aircraft may increase the frequency of contrail formation, their climate impact is partly offset by the absence of soot and other particle emissions, which results in fewer but larger ice crystals with lower optical depth and a shorter lifetime (Tiwari et al. 2024, Voigt 2021). Keeping in view the possibilities of great environmental benefits of hydrogen based aircraft, it becomes most competitive clean energy alternative. However, the stakeholders must bring in the niche innovation to solve the technical challenges of this fuel so that it disrupts the existing regime.

3.4 Battery Electric Aircraft (BEA)

Fully battery-electric aircraft offer clear advantages, including zero in-flight emissions, lower noise levels and reduced operating costs. At the same time, they present major design and engineering challenges. This section reviews current progress in battery-electric aircraft and considers the economic and environmental issues linked to their development. In a fully electric aircraft, rechargeable batteries supply power to an electric motor, which in turn drives the propeller. At present, this form of propulsion is largely limited to short-range light aircraft, including eVTOL vehicles, small conventional fixed-wing aircraft and newer fixed-wing concepts (Adu-Gyamfi and Good 2022). Publicly announced examples include the Pipistrel Velis Electro, Beta Technologies' Alia-250, Eviation's Alice and the Lilium Jet (Adu-Gyamfi and Good 2022). Even so, the main barrier to wider adoption remains the limited range of these aircraft, which is largely a result of the low energy density of current battery technology.

Caset et al. (2018) note that, with present battery technology, an Airbus A320 would require a battery weighing around 38 times its maximum take-off weight. For battery-electric aircraft to become viable in general aviation and regional operations, battery energy density would need to reach at least 400 Wh/kg and 750 Wh/kg respectively (Lvovich and Lawson 2018). This shows that progress in battery-electric aviation depends heavily on advances in battery performance, more efficient electric motors and improved airframe design.

Adu-Gyamfi et al. (2022) also compare electric propulsion with conventional regional aircraft powered by jet fuel as depicted in Figure 8. The most advanced lithium-ion batteries currently offer an energy density of about 250 Wh/kg, whereas Jet A-1 fuel provides around 12,000 Wh/kg, making it far more energy dense. However, this gap is partly reduced by the much higher efficiency of electric motors. While the fuel-to-wheel efficiency of an internal combustion engine is about 28%, an electric motor can reach around 90% efficiency, reducing the effective energy density gap from 1:48 to about 1:15 (Ahman 2001). Adu-Gyamfi et al. (2022) argue that if battery energy density rises to 600 Wh/kg, this ratio could fall further to 1:6.2. They also suggest that aerodynamic improvements, such as distributed propulsion and boundary layer ingestion, could cut total in-flight energy demand by a factor of three to five, bringing the requirement closer to 1:1.5. Taken together, this suggests that making battery-electric aircraft a practical reality will require progress in three areas: higher battery energy density, better electric motor performance and more efficient aerodynamic design.

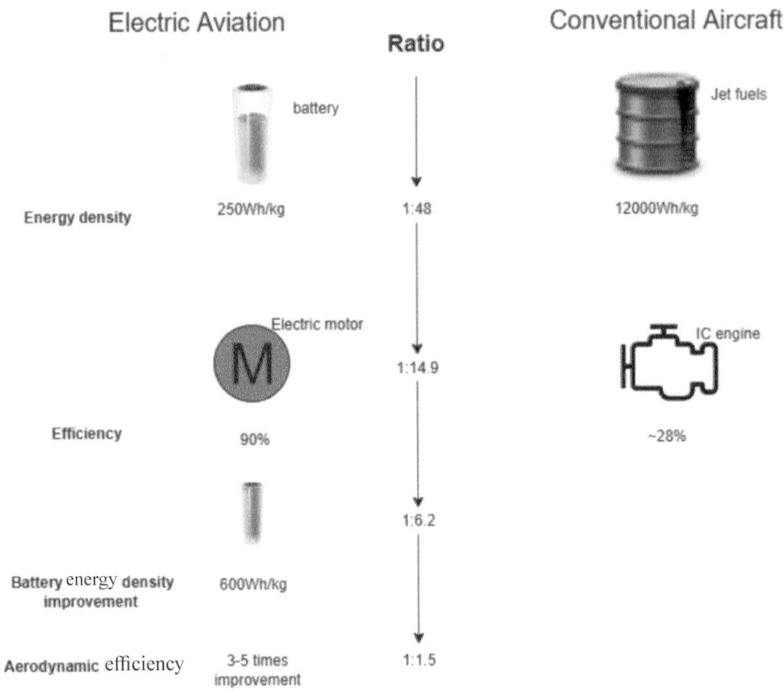

Figure 8. Comparative features of electric propulsion to conventional jet fuel regional aircrafts. Source: (Adopted from Adu-Gyamfi et al. 2022).

The Table 4 in appendix summarises the list of electric aircraft concepts. It is observed that there are only five concepts available in literature that are based on fully electric aircraft. This includes VoltAir designed by Airbus which has seating capacity of 68 requiring battery density of 1000 Wh/kg. X57 is fully electric aircraft that has been designed by NASA and powered by Li-ion battery with a range of 87 nm and seating capacity of 2 passengers. Similarly, Luftfahrt Celiner developed by Bauhaus requires battery energy density of 2000 Wh/kg for a seat count of 189 and range of 900 nm. Another concept has been led by Project Pegasus from NASA, which is a full electric aircraft with 48 seats and a range of 200 nm requiring battery energy density of 750–1000 Wh/kg. In addition, the Alice being developed by Eviation aims to design 9-seater aircraft with a range of 440 nm. The TRL of BEA are low, a few small range aircraft may start operating by 2040, however large scale transition to BEA is far off.

It is evident that the BEA technology is still fledgling and gradually maturing. It's difficult to understand the BEA costs due to uncertainty of the technology. Avogadro et al. (2024) have evaluated the cost associated for the First Generation Electric Aircraft (FGEA) with the incumbent aircrafts. His study identifies that FGEA show 25–30% higher ownership cost than the conventional counterparts. The crew costs are 15–25% higher than conventional aircrafts. Fuel costs are 75% lower while maintenance cost is predicted to be lower but value remains undetermined.

This study suggests that overall, FGEA have higher operating cost in range of 10–30% than that of incumbent aircrafts.

4. Role of Socio-technical Approach to Accelerate the Clean Energy Transition in Aviation

The net zero emission goals in aviation cannot be achieved by technological solutions only. The transition to low carbon trajectory involves multi-dimensional and complex interactions between industry technology, market, policy, society and culture (Geels 2002). A successful adoption of low carbon technology solutions will need effective collaborations between producers, suppliers, airlines, government policies and societal users and other stakeholders. Such collaborations are complex and inter-related encompassing social, economic, political perspectives. Hence, the techno-economic perspectives are not sufficient to plan for system transition pathways. Given this, the inclusion of social insights adds a meaningful insight in sustainable energy transitions (Lennon et al. 2019). Geels (2002) propounded the Multi-Level Perspective (MLP) which is widely recognised framework for socio-technical analysis. The transitions are considered to be non-linear processes evolving from interplay of developments occurring at three analytical level shown in Figure 9.

1. The changes driven by technology and external factors exerts pressure on the socio-technical landscape.
2. This exogenous pressure destabilizes the current regime creating new opportunities and misalignment in existing activities.
3. Innovation within niche accelerates the transition to a stabilized new regime.

Urban et al. (2024) utilised MLP approach to evaluate the sustainable transition of the aviation sector. It was observed that sustainability transition of aviation is slowly changing due to landscape pressure such as Paris agreement, ICAO's CORSIA, IATA's Fly net zero and EU fit for 55 policy. Several technology trajectories such as SAF, Hydrogen fuelled & Electric aircrafts are being pursued by the system actors. These technologies call for system-wide change across infrastructure, supply chains, aircraft design, business models and industry partnerships, giving them the potential to disrupt the current regime and support sustainable transitions; however, Urban et al. (2024) show that incumbent actors still play a central role, with niche innovations often emerging through collaboration between established and new entrants.

Muehlberger et al. (2024) has reviewed the strategies suggested in six major decarbonisation reports and have synthesised the three socio-technical approaches suggested in the reports, which are summarised in Table 5 in appendix:

a. Rethinking travel and behaviour change,
b. Radical modernisation and technological and
c. Transition to renewable fuel and clean energy sources

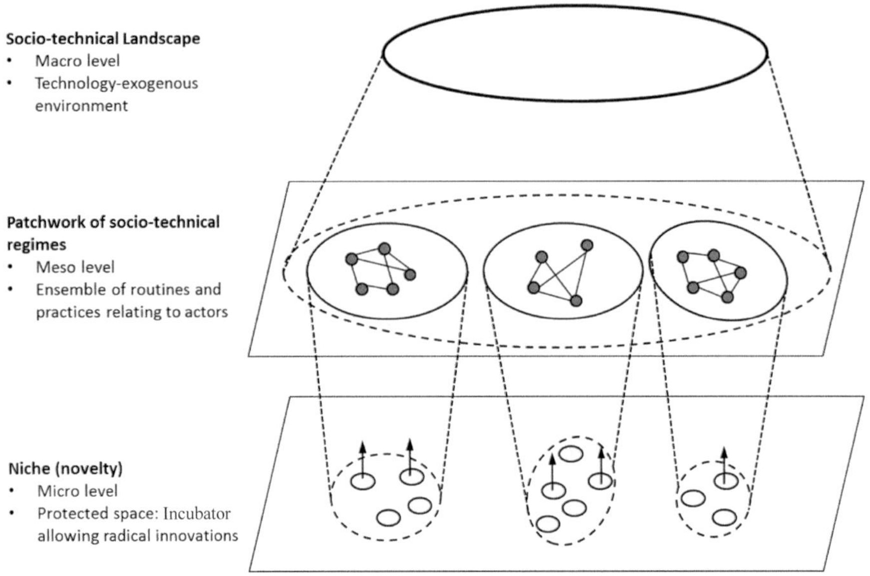

Figure 9. Multiple-level nested hierarchy. Source: (Adopted from Kim et al. 2019).

The urgency of transitioning to sustainable energy is widely recognized; however, the strategies for achieving this shift remain highly political, raising the question of whether existing socio-technical systems should be maintained or restructured. Stakeholders from research, industry, and consulting sectors advocate for green growth, promoting advanced technologies such as hydrogen-powered aviation to sustain economic progress. On the other hand, environmental groups support the concept of transport sufficiency, emphasizing the need to reduce demand in the aviation sector. This debate also extends to whether the fundamental structures of the socio-technical system should be preserved or transformed.

Among the three proposed approaches in the study of Muehlberger et al. (2024), fuel innovation is the most compatible with maintaining the current system, as it enables the continued use of existing fuel infrastructure, airport facilities, and user behaviours. In contrast, fleet innovation requires substantial changes across the aviation supply chain, while travel innovation promotes shifts in user practices, both contributing to system-wide restructuring (Muehlberger et al. 2024). As a result, fuel innovation could slow down a more profound socio-technical transition within the aviation industry. Given its ease of integration into the existing system, fuel innovation appears to be the most likely to materialize, though its implementation remains uncertain. Political actors and other interest groups are expected to play a key role in shaping this transition (Muehlberger et al. 2024). It has been evident that policy innovations put pressure on incumbent regime to bring regime shift. An excellent example of implementation of policy innovation is the from Denmark where government enacted two important legislations based on socio-technical approach: a. promotion of Renewable Energy, b. Danish electricity supply act which accelerated the energy transition towards renewable energy bringing value to the society (Leal-Arcas 2023). A similar policy innovation will accelerate the renewable energy transition in air transport sector.

5. Conclusion

Over the last two decades, GHG emissions from the aviation sector have increased significantly and without restraint. There is growing interests and momentum in the industry to reduce emissions which are witnessed by ICAO's LTAG and IATA's Fly net zero by 2050 commitments at strategic level. Although it has been observed in the chapter that the rate of the growth of the sector outpaces the decarbonisation efforts.

Currently, SAF is the only clean technology solution that is ready for deployment in the sector. However, due to low production scale, high production cost and infant supply chain of SAF feedstock, the uptake of SAF still remains low. There is urgent need for the sector to find a solution to accelerate the SAF uptake in the industry. The solution to accelerate the implementation of this technology could be based on the socio-technical approach. There is need for all actors of the sector to identify the socio-technical instruments which could help to rapidly deploy the technology.

Although it is observed in the chapter that the SAF on its own is not sufficient to achieve the net zero emission targets. The industry needs to find the clean energy technology options which could eliminate the emissions of CO_2 and significantly reduce the non-CO_2 emissions as well. Hydrogen fuelled or Battery Electric aircraft are the potential alternatives of future clean technology which could address these challenges. However, the key technological challenge for hydrogen fuelled aircrafts are storing and handling hydrogen on board. Fuel options such as compressed gas, liquefied hydrogen, and fuel cells are being evaluated to mitigate hydrogen's weight and drag penalties. Similarly for Battery Electric Aircraft, the key technological challenge is to enhance the low energy density of current battery technology. Achieving the battery density threshold of 1000 Wh/kg could accelerate the development of this technology option.

From the analysis it is identified that the technological innovation on its own is not sufficient to create the transition pathways. The adoption of clean technology in aviation will require a systems' changes such as new production and distribution facilities, infrastructure at airport, new supply chains, changes in aircraft designs, business models and alliances. Apart from technological innovation, there is also a need to develop the solutions from socio-technological approach so that the clean technology transitions could be accelerated. The possibility of collective socio-technical levers to accelerate the clean energy transition in the aviation sector should be examined.

References

Abrantes, I., Ferreira, A. F., Silva, A. and Costa, M. 2021. Sustainable aviation fuels and imminent technologies-CO_2 emissions evolution towards 2050. Journal of Cleaner Production 313: 127937.

Abram, S., Atkins, E., Dietzel, A., Jenkins, K., Kiamba, L., Kirshner, J. et al. 2022. Just Transition: A whole-systems approach to decarbonisation. Climate Policy 22(8): 1033–1049.

Adler, E. J. and Martins, J. R. 2023. Hydrogen-powered aircraft: Fundamental concepts, key technologies, and environmental impacts. Progress in Aerospace Sciences 141: 100922.

Adu-Gyamfi, B. A. and Good, C. 2022. Electric aviation: A review of concepts and enabling technologies. Transportation Engineering 9: 100134.

Agusdinata, D. B. and Delaurentis, D. A. 2011. Addressing equity issue in multi-actor policymaking via a system-of-systems approach: Aviation emissions reduction case study. Journal of Systems Science and Systems Engineering 20: 1–24.

Åhman, M. 2001. Primary energy efficiency of alternative powertrains in vehicles. Energy 26(11): 973–989.

ATAG. 2021. Waypoint 2050 Report 2nd Edition. https://atag.org/resources/waypoint-2050-2nd-edition-september-2021.

ATI. 2022. Destination Zero-The technology journey to 2050. https://www.ati.org.uk/wp-content/uploads/2022/04/ATI-Tech-Strategy-2022-Destination-Zero.pdf.

Åkerman, J., Kamb, A., Larsson, J. and Nässén, J. 2021. Low-carbon scenarios for long-distance travel 2060. Transportation Research Part D: Transport and Environment 99: 103010.

Avogadro, N. and Redondi, R. 2024. Demystifying electric aircraft's role in aviation decarbonization: Are first-generation electric aircraft cost-effective?. Transportation Research Part D: Transport and Environment 130: 104191.

Braun, M., Grimme, W. and Oesingmann, K. 2024. Pathway to net zero: Reviewing sustainable aviation fuels, environmental impacts and pricing. Journal of Air Transport Management 117: 102580.

Brewer, G. D., Morris, R. E., Davis, G. W., Versaw, E. F., Cunnington, G. R., Riple, J. C. et al. 1978. Study of fuel systems for LH2-fueled subsonic transport aircraft, Contractor report NASA/CR–145369, NASA. URL https: //ntrs.nasa.gov/citations/19780023142.

Brewer, G. D. 2017. Hydrogen Aircraft Technology. Routledge.

Bergero, C., Gosnell, G., Gielen, D., Kang, S., Bazilian, M. and Davis, S. J. 2023. Pathways to net-zero emissions from aviation. Nature Sustainability 6(4): 404–414.

Braun, M., Grimme, W. and Oesingmann, K. 2024. Pathway to net zero: Reviewing sustainable aviation fuels, environmental impacts and pricing. Journal of Air Transport Management 117: 102580.

Caset, F., Boussauw, K. and Storme, T. 2018. Meet & fly: Sustainable transport academics and the elephant in the room. Journal of Transport Geography 70: 64–67.

Dahal, K., Brynolf, S., Xisto, C., Hansson, J., Grahn, M., Grönstedt, T. et al. 2021. Techno-economic review of alternative fuels and propulsion systems for the aviation sector. Renewable and Sustainable Energy Reviews 151: 111564.

Department for Transport, 2023. https://www.gov.uk/government/news/lift-off-for-projects-fuelling-jet-liners-with-bin-liners.

Doliente, S. S., Narayan, A., Tapia, J. F. D., Samsatli, N. J., Zhao, Y. and Samsatli, S. 2020. Bio-aviation fuel: a comprehensive review and analysis of the supply chain components. Frontiers in Energy Research 8: 499009.

Dray, L., Schäfer, A. W., Grobler, C., Falter, C., Allroggen, F., Stettler, M. E. et al. 2022. Cost and emissions pathways towards net-zero climate impacts in aviation. Nature Climate Change 12(10): 956–962.

EASA. 2024. EASA supporting scale up of SAF. https://www.easa.europa.eu/en/newsroom-and-events/press-releases/easa-supporting-scale-sustainable-aviation-fuels-through-eu.

Eswaran, S., Subramaniam, S., Geleynse, S., Brandt, K., Wolcott, M. and Zhang, X. 2021. Techno-economic analysis of catalytic hydrothermolysis pathway for jet fuel production. Renewable and Sustainable Energy Reviews 151: 111516.

Eurocontrol. 2022. Objective Sky Green. https://www.eurocontrol.int/sites/default/files/2022-05/eurocontrol-objective-skygreen-2022-2030-report-20220523.pdf.

European Commission. 2023. https://transport.ec.europa.eu/transport-modes/air/environment/refueleu-aviation_en.

Geels, F. W. 2005a. Processes and patterns in transitions and system innovations: Refining the co-evolutionary multi-level perspective. Technological Forecasting and Social Change 72(6): 681–696.

Geels, F. W. 2005b. Technological Transitions and System Innovations: A Co-Evolutionary and Socio-Technical Analysis. Edward Elgar Publishing.

Geels, F. W. 2012. A socio-technical analysis of low-carbon transitions: introducing the multi-level perspective into transport studies. Journal of Transport Geography 24: 471–482.

Goh, B. H. H., Chong, C. T., Ong, H. C., Seljak, T., Katrašnik, T., Józsa, V. et al. 2022. Recent advancements in catalytic conversion pathways for synthetic jet fuel produced from bioresources. Energy Conversion and Management 251: 114974.

Gössling, S. and Lyle, C. 2021. Transition policies for climatically sustainable aviation. Transport Reviews 41(5): 643–658.

Gössling, S. and Humpe, A. 2023. Net-zero aviation: Time for a new business model?. Journal of Air Transport Management 107: 102353.

Gössling, S., Humpe, A. and Sun, Y. Y. 2024. On track to net-zero? Large tourism enterprises and climate change. Tourism Management 100: 104842.

Gössling, S., Humpe, A. and Sun, Y. Y. 2025. Are emissions from global air transport significantly underestimated?. Current Issues in Tourism 28(5): 695–708.

Hoelzen, J., Silberhorn, D., Zill, T., Bensmann, B. and Hanke-Rauschenbach, R. 2022. Hydrogen-powered aviation and its reliance on green hydrogen infrastructure—Review and research gaps. International Journal of Hydrogen Energy 47(5): 3108–3130.

IATA. 2021. https://www.iata.org/en/programs/sustainability/flynetzero/.

ICAO. 2019. CORSIA. https://www.icao.int/environmental-protection/CORSIA/Pages/default.aspx.

ICAO. 2022. Long term aspirational goal for 2050. https://www.icao.int/environmental-protection/Pages/LTAG.aspx.

ICCT. 2023. https://theicct.org/publication/global-aviation-icao-net-zero-goal-jan23/.

IPCC. 2023. Climate Change Report for Policy Makers. https://www.ipcc.ch/report/ar6/syr/downloads/report/IPCC_AR6_SYR_SPM.pdf.

International Energy Agency (IEA). 2019. The Future of Hydrogen, https://www.iea.org/reports/the-future-of-hydrogen.

IEA. 2020. Tracking Transport 2020. International Energy Agency, Paris, France.

IEA. 2023. Tracking Clean Energy Progress 2023, IEA. Paris. https://www.iea.org/reports/tracking-clean-energy-progress-2023, Licence: CC BY 4.0.

IPCC. 2020. Mitigation of Climate Change 6th Report. https://www.ipcc.ch/report/ar6/wg3/chapter/chapter-10/#:~:text=Global%20transport%20was%20the%20fourth,2%20emissions%20(IEA%20 2020a).

Kim, Y., Lee, J. and Ahn, J. 2019. Innovation towards sustainable technologies: A socio-technical perspective on accelerating transition to aviation biofuel. Technological Forecasting and Social Change 145: 317–329.

Leal-Arcas, R. 2023. Electricity Decentralization in The European Union: Towards Zero Carbon and Energy Transition. Elsevier.

Lennon, B., Dunphy, N. P. and Sanvicente, E. 2019. Community acceptability and the energy transition: A citizens' perspective. Energy, Sustainability and Society 9(1): 1–18.

Lee, D. S., Fahey, D. W., Skowron, A., Allen, M. R., Burkhardt, U., Chen, Q. et al. 2021. The contribution of global aviation to anthropogenic climate forcing for 2000 to 2018. Atmospheric Environment 244: 117834.

Lvovich, V. and Lawson, J. 2018, September. Integrated computational-experimental development of lithium-air batteries for electric aircraft. In Annual Meeting of International Society of Electrochemistry (No. GRC-E-DAA-TN60215).

Mukhopadhaya, J. and Rutherford, D. 2022. Performance analysis of evolutionary hydrogen-powered aircraft. International Council on Clean Transportation.

Muehlberger, C. M., Gruen, L., Liefner, I. and Losacker, S. 2024. Socio-technical imaginaries of climate-neutral aviation. Energy Research & Social Science 114: 103595.

O'hayre, R., Cha, S. W., Colella, W. and Prinz, F. B. 2016. Fuel Cell Fundamentals. John Wiley & Sons.

Paris Agreement. 2015. https://unfccc.int/process-and-meetings/the-paris-agreement.

Park, H., Chae, H. J., Suh, Y. W., Chung, Y. M. and Park, M. J. 2022. Techno-economic analysis and CO_2 emissions of the bioethanol-to-jet fuel process. ACS Sustainable Chemistry & Engineering 10(36): 12016–12022.

Pausas, J. G. and Keeley, J. E. 2021. Wildfires and global change. Frontiers in Ecology and the Environment 19(7): 387–395.

Ravichandran, D. 2024. https://etn.news/hydrogen/airbus-hydrogen-aircraft-combustion-fuel-cell-tech-validations.

Scheelhaase, J. and Maertens, S. 2020. How to improve the global 'carbon offsetting and reduction scheme for international aviation'(CORSIA)?. Transportation Research Procedia 51: 108–117.

Tiwari, S., Pekris, M. J. and Doherty, J. J. 2024. A review of liquid hydrogen aircraft and propulsion technologies. International Journal of Hydrogen Energy 57: 1174–1196.

Urban, F., Nurdiawati, A., Harahap, F. and Morozovska, K. 2024. Decarbonizing maritime shipping and aviation: Disruption, regime resistance and breaking through carbon lock-in and path dependency in hard-to-abate transport sectors. Environmental Innovation and Societal Transitions 52: 100854.

U.S. Department of Energy. 2021. Hydrogen shot. https://www.energy.gov/ eere/fuelcells/hydrogen-shot.

Verstraete, D. 2013. Long range transport aircraft using hydrogen fuel. International Journal of Hydrogen Energy 38(34): 14824–14831.

Virgin Atlantic. 2023. World's first 100% SAF fuelled commercial aircraft. https://corporate.virginatlantic.com/gb/en/media/press-releases/worlds-first-sustainable-aviation-fuel-flight.html.

Voigt, C., Kleine, J., Sauer, D., Moore, R. H., Bräuer, T., Le Clercq, P. et al. 2021. Cleaner burning aviation fuels can reduce contrail cloudiness. Communications Earth & Environment 2(1): 114.

Warnecke, C., Schneider, L., Day, T., La Hoz Theuer, S. and Fearnehough, H. 2019. Robust eligibility criteria essential for new global scheme to offset aviation emissions. Nature Climate Change 9(3): 218–221.

Watson, M. J., Machado, P. G., da Silva, A. V., Saltar, Y., Ribeiro, C. O., Nascimento, C. A. O. et al. 2024. Sustainable aviation fuel technologies, costs, emissions, policies, and markets: a critical review. Journal of Cleaner Production 449: 141472.

Zhou, Y., Searle, S. and Baldino, C. 2022. Cost of renewable hydrogen produced onsite at hydrogen refueling stations in Europe. International Council on Clean Transportation: Washington, DC, USA, p.17.

Appendix

List of Tables

Table 1. Comparative cost benefits of SAF pathways with conventional jet fuel.

SAF Production Pathways	MJSP average per litre in USD	Percentage price increase from Jet A1	CO_2 emissions average g CO_2 eq/MJ	Percentage decrease from Jet A1
FT-SPK	2.08	320%	7.84	92%
HEFA-SPK	1.12	120%	65.22	27%
ATJ-SPK	1.69	240%	38.67	57%
HFS-SIP	3.99	700%	42.02	53%
CHJ-SPK	1.3	160%	20.58	77%

Source: (Data adopted from Watson et al. 2024)

Table 2. Comparison of characteristics of Jet fuel with hydrogen fuel.

Characteristics	Conventional Jet Fuel	Liquid Hydrogen	Gaseous Hydrogen (350 bar)	Gaseous Hydrogen (700 bar)
Specific energy (MJ/kg)	43.2	120	120	120
Energy density (MJ/L)	34.9	8.5	2.9	4.8
Storage temperature (K)	Ambient	20	Ambient	Ambient
Storage pressure (bar)	Ambient	2	350	700
Tank gravimetric efficiency (%)	100	30 to 90	1 to 15	1 to 15

Source: (Data adopted from Adler et al. 2023)

Table 3. Current hydrogen powered aircraft and their storage and propulsion technology.

Aircraft	First Flight	Storage	Propulsion
NASA modified B-57	1957	LH_2	Turbojet
Tupolev Tu-155	1988	LH_2	Turbofan
Boeing Fuel Cell Demonstrator Airplane	2008	GH_2	PEMFC
Antares DLR H2	2009	GH_2, 350 bar	33 kW fuel cell
AeroVironment Global Observer	2011	LH_2	
Boeing Phantom Eye	2012	LH_2	Modified Ford 2.3 L ICE
H2FLY HY4	2016	GH_2	45 kW PEMFC
ZeroAvia Piper Malibu demonstrator	2020	GH_2, 350 bar	PEMFC
ZeroAvia Dornier 228 demonstrator	2023	GH_2	Fuel cell
Universal Hydrogen Dash-8 demonstrator	2023	GH_2	PEMFC

Source: (Data adopted from Adler et al. 2023)

Table 4. Current electric aircraft proposed concepts.

Aircraft Name	Organization	Time frame	E-Architecture	Seat Count	Design Range (nm)	Airframe propulsion architecture	Components	Emissions/Aircraft Performance
SAR-40	Cambridge-MIT Institute	2030	TE	215	5000	BLI		Achieved low noise level below 65dBA
Double bubble-D8	NASA/MIT	2035	ICE	180	3000	BLI		8.5% reduction in fuel burn
E-Thrust	Airbus/Rolls Royce	2050	SHE	90	Regional	DP/BLI	1000Wh/kg Li-air batteries, HTS motors	A high effective bypass ratio
VoltAir	EADS/Airbus	2035	E	68	900	BLI	1000Wh/kg Li-air batteries, HTS motors	0% CO_2 emission, 25% energy efficient
N3-X	NASA	2045	TE	300	7500	DP/BLI	HTS technology	70% reduction in fuel burn
Sugar Volt	Boeing	2035	PHE	154	3500		Battery-specific energy>750 Wh/kg	60% reduction in fuel burn, 80% reduction in Nox
X-57 "Maxwell"	NASA	2019	E	2	87	DEP	Li-ion batteries, 14 electric motors	500% increase in high speed cruise efficiency, Zero in-flight emissions
STARC-ABL	NASA	2035	TE	154	3500	BLI	Electric motors/generators	12% reduction in fuel burn
Luftfahrt Ce-Liner	Bauhaus	2035	E	189	900	DEP with ducted fans	2000Wh/kg battery, HTS technology	Zero inflight emissions and improved aerodynamics
PEGASUS	NASA	2030	E, PHE	48	200, 400	DEP	750-1000 Wh/kg battery, HTS technology	Upto 40% energy efficiency
Alice	Eviation	2019	E	9	440	BLI	No future technology implemented	
ECO-150	ESAero	2035	TE	150	1650	DP	8 HTS electric motors and turbo electric generators	40% reduction in fuel burn
ZA10	Zunum Aero	2020s	SHE	12	700		Conventional regional jet layout with a V-tail	80% reduction in CO_2 and noise emission

Source: (Data adopted from Adu-Gyamfi et al. 2022)

Table 5. Socio-technical themes, characteristics, challenges, activities and stakeholders.

Socio-technical themes	**Characteristics**	**Challenges**	**Main activities**	**Economic trajectory**	**Stakeholders**
Rethinking travel and behavioural change (travel innovation)	Demand reduction, behavioural change, no sole dependence on technology, economic measures	Legal & political barriers, capping traffic levels, government regulations	Reduce corporate travel, carbon pricing, stop aviation subsidies and airport expansion	Demand management	Environmental Associations
Radical modernization and technological progress (fleet innovation)	Disruptive technologies novelties such as H2 aircraft or electric aircraft. SAF depolyment as short run remedies.	Huge investment in research and innovation. Infrastructure required. Financial risks. Certification challenges.	Investment needed in new technologies aircrafts	Ecological modernization	Research and technology, consulting, Industry
Transition to alternative fuels and renewable energy sources	Transition to low carbon fuel or carbon free fuels, aggressive SAF scale-up, development of renewable energy sources.	Huge investment in SAF production. High Investment risks. Availability of feedstocks, price gap between fossil fuels and SAF	Invest in renewable energy sources, incentivize SAF use, reduce cost of SAF production	Business-as-usual	Cross-sector think tank (Industry)

Source: (Data adopted from Muehlberger et al. 2024)

Chapter 7

Navigating Complex Challenges

Examining Barriers for Attaining Energy Security in Emerging Economies

Juhi Raghuvanshi[1,*] and *Aakash Khindri*[2]

1. Introduction

Human civilization has become more complicated, necessitating the reliable and varied supply of energy sources in order to maintain its operations (Smil 2007, Armaroli and Balzani 2007). We first extract raw materials and transform them into secondary products, and then refine or develop commodities originating from these raw materials. Some varied functions like powering vehicles and heating structures, energizing industrial gear, and facilitating the manufacturing of less used consumer items are the secondary products which are very crucial in contemporary life. Securing a dependable supply of these resources is essential for a nation's overall well-being (Ibekwe 2024).

With time, there is a major increasement in global trade volumes in contrast to previous times when remote communities contributed in limited exchanges. Recently, most traded commodities are refined petroleum oils, petroleum gases, and crude oil. Despite progress in energy alternatives, nations remain extremely reliant on fossil fuels, depict energy security a pivotal geopolitical concern. As supply disruptions can precipitate political and social turmoil, energy security

[1] Innovation, Entrepreneurship, and Venture (IEV) Department at Netaji Subhas University of Technology (NSUT), New Delhi.

[2] Jindal Global Business School, O.P. Jindal Global University, Sonipat, Haryana.

* Corresponding author: raghuvanshijuhi8@gmail.com

DOI: 10.1201/9781003475385-7

influences directly to national stability. Therefore, it has a vital role in economic growth, affecting industrial productivity, trade competitiveness, and energy pricing. Moreover, international relations, influencing alliances, conflicts, and power dynamics among nations are essential in energy resources. Energy security is fundamental to sustainable development and international stability but not merely a technical or business concern.

Energy security evolved as a policy issue in the early twentieth century as a result of the need to supply oil to armies (Fukutomi 2024). Academic debates on energy security has begun in the 1960s (Lubell 1961) and culminated in the 1970s with the oil crisis. Academic interest in energy security waned in the late 1980s and 1990s, as oil prices stabilized and the fear of political embargoes faded. It resurfaced in the 2000s, owing to increased demand in Asia, disruptions in European gas supplies, and efforts to decarbonize the energy system (Yergin 2006, Hancock and Vivoda 2014).

High oil prices in the years leading up to 2008, as well as geopolitical supply problems, have reignited interest in energy security (IEA 2007a, b). The recent high oil prices can be attributed to a variety of factors. The most widely accepted interpretation is that they reflect rapidly rising Asian energy demand, underinvestment in energy supply in the late 1990s and early 2000s, and increased concentration of oil and gas deposits in a few politically unstable countries (Van der Linde 2004). Other sources, on the other hand, link the current spike in oil prices to the depletion of (cheap) oil resources. Governments all throughout the world have responded to the current situation by enacting policies to increase supply security. However, in the vast majority of cases, this aspiration is not expressed in concrete objectives. This is due in part to the fact that energy security is a nebulous concept that is very context dependent. Despite this, governments continue to priorities supply security as a top priority in their energy policies. Since energy security is deeply interconnected with other energy policy concerns (such as environmental policy, climate change and inexpensive energy), It is important to assess how different growth pathways impact energy security. This requires clearly defined indicators of energy security, as well as a systematic framework that links the concept with model based scenario analysis.

Taking this into consideration, the present study identifies barriers to energy security and assists locally owned small-scale renewable energy projects in addressing the most significant barriers first. This study proposes a robust model based on the DEMATEL technique to rank the identified barriers to energy security. Finally, a sensitivity analysis is performed to demonstrate the model's robustness.

The chapter is structured into seven sections. The first section gives an introduction, followed by a literature review in the second section. The third section outlines the research methodology employed in the study. The fourth section presents the results and analysis, while the fifth section discusses their implications. The sixth section concludes the chapter, and the seventh section highlights directions for future research.

1.1 Energy in Remote and the Rural Communities of India

Remote and rural communities in India face substantial challenges with insufficient and unreliable energy access, leading to frequent outages that disrupt daily life and economic activities. A significant portion of rural households, especially in remote areas, remain energy-poor, heavily relying on traditional biomass like firewood for cooking and heating, which raises health and environmental concerns. The distribution inefficiencies, and frequent power outages hinder fundamental services such as healthcare, education, and agriculture, impacting productivity and quality of life because of lack of dependable grid-based electricity. Women and children bear the burden of gathering traditional fuels, while students struggle to study after dark.

Despite initiatives by the Indian government, such as the Deendayal Upadhyaya Gram Jyoti Yojana and the Saubhagya Scheme, continuous and reliable energy access remains a challenge. No study to date has thoroughly identified and prioritizing the barriers to the energy security in the emerging economies. This study utilized a robust model utilizing the Decision Making Trial and Evaluation Laboratory (DEMATEL) to identify the most significant barriers to Energy Security (ES) that demand immediate attention.

2. Factors Identification from the Literature

2.1 Community Resistance

Cultural attachment with the traditions, increase the community resistance to adopt the novel energy technologies which further enhance the unintentional resistance to use modern energy solutions (Barnett 2000). The execution and the utilization of novel technologies is obstructed by this resistance. Hence, the preservation and accomplishment are completely depending upon the support from the community (Sovacool and Dworkin 2015). Moreover, the cost associated with the implementation of the project further increased, as there is a necessity for creating awareness via outreach programme regarding the importance of such projects in the community (Qazi et al. 2021).

2.2 Limited Decision-Making Capability

People in the community lack the confidence and the autonomy to make the substantial decisions concerning the initiatives regarding energy, which delays the execution and this slows down the adoption (Matinga and Annegarn 2013). Native's opinion is not recognized sue to the external influence and partial exposure regarding the processes of energy projects, which significantly affects the decision making capabilities. Lack of organization has an adverse effect on the investment and the commitment by the community for energy solution, which results in lost prospects for sustainable development (Mandelli et al. 2016, Luthra et al. 2016).

2.3 Insufficient Community Involvement

Communities are marginalized from essential decision-making processes due to insufficient community involvement in energy planning, potentially resulting in energy solutions that do not align with their needs or beliefs (Reed 2008). This inconsistency jeopardizes the energy security goals and reduces the likelihood of successful project implementation. Inclusive participation ensures that energy projects address local needs, hence fostering sustainability and ownership (Veelen 2018).

2.4 Low Financial and Energy Awareness

The capability of the communities in the rural area about the energy services is often hindered due to insufficient energy and financial literacy (Lee et al. 2019). This gap hinders effective participation in or access to energy initiatives, hence limiting the broader goal of energy security. To empower communities to invest in and utilise energy responsibly, it is essential to improve financial and energy literacy (Sovacool 2012, Pachauri and Spreng 2012).

2.5 Limited Knowledge of Energy Conservation and Efficiency

Communities in the rural area faces power cuts and high costs due to inefficient usage of energy due to the lack of energy conservation awareness (Reddy and Balachandra 2006). Judicious management of the resources requires the understanding and knowledge of the strategies related to the energy saving (UNEP 2011). Awareness regarding the energy effective usage and management also lessen the weight on current infrastructure thus helps in improving the energy security (Modi et al. 2005, Ramachandra 2009, Chaurey and Kandpal 2010).

2.6 Undervaluing Traditional Knowledge Integration

Undervaluing the integration of traditional knowledge and the local practices sometimes results in culturally incompatible and environmentally unsustainable energy initiatives. Traditional knowledge provides valuable comprehensions into energy usage which advances the project resilience and community acceptance. Additionally, context-specific solutions can be achieved by recognizing the traditional knowledge as an asset (Berkes 2012).

2.7 Limited Project Development Experience

The success of the initiatives taken in the rural communities are depending upon the experience of the project handling team, which includes proper planning, resource allocation and managing risk (Wlokas et al. 2012). Improper design, improper funds allocations and unmet objectives are the consequences of lack of relevant experiences which further expose the issues related to energy security (Ahlborg and Hammar 2014). These difficulties can be dodged through the execution of an effective project management framework (Chaurey et al. 2012, Singh 2021).

2.8 Insufficient Expertise in Energy-Efficient Programs

It is important to own expertise in technologies related to energy-efficient for reducing the costs and energy consumption optimization (Janda and Parag 2013). However, rural community often experience a lack of access to programs related to energy efficiency and knowledge, resulting in systems that are both inefficient and expensive. The stress on energy resources is exacerbated, compromising energy security and reliability, in the absence of programs that priorities energy efficiency (Sovacool 2012, Bhattacharyya and Palit 2014).

The results of the aforementioned studies indicate that researchers have made an effort to identify the barriers of energy security barriers, albeit in a dispersed way (refer to Table 1). Prominent scholars in the field contend that additional research is required to present a comprehensive understanding of the causal relationships among energy security barriers, thereby prioritizing and mapping these connections, and to move beyond isolated elements. To the author's best understanding, no previous study has provided a comprehensive understanding of the connections between the various causal barriers of energy security and their outcome barriers, where one barrier may potentially contribute to another. Consequently, this investigation has identified energy security barriers from the existing literature in order to address this lacuna. Recent research has revealed that numerous of these barriers are prevalent worldwide and are frequently investigated in conjunction with other related factors. A comprehensive literature review of variables associated with energy security barriers is provided in Table 1.

Table 1. List of the identified energy security barriers.

S.No.	Code	Barriers
1	ES1	Community Resistance
2	ES2	Insufficient Community Involvement
3	ES3	Limited Knowledge of Energy Conservation and Efficiency
4	ES4	Low Financial and Energy Awareness
5	ES5	Limited Decision-Making Capability
6	ES6	Undervaluing Traditional Knowledge Integration
7	ES7	Limited Project Development Experience
8	ES8	Insufficient Expertise in Energy-Efficient Programs

Source: Author's Own

3. Methodology

3.1 DEMATEL

The aim of this research is to determine whether any energy security barrier has a direct effect on another, thereby examining the potential for one factor to lead to another. Comprehending the causal relationships between barriers that generate another causal barrier. The DEMATEL technique has been implemented by numerous distinguished authors (e.g., Raghuvanshi et al. 2017, Chakrabotry et al. 2018, Raghuvanshi and Agrawal 2020, Ranjan et al. 2016) to investigate intricate causal relationships among a variety of barriers and obstacles. The present study employs the DEMATEL technique, which is also beneficial for the analysis of intricate relationships through graph theory, in accordance with a comparable methodology (Zhou et al. 2011). DEMATEL, which was created by the Battelle Memorial Institute in Geneva, is intended to elucidate complex situations that involve interconnected factors. The following progressive stages are taken in this study to establish causal relationships among energy security barriers in small-scale energy projects, in accordance with a methodology similar to that of Raghuvanshi et al. (2017).

Step I: Develop a Direct relationship matrices (M): Five respondents were requested to construct a direct relationship matrix among the factors/variables enumerated in Table 2 (for one respondent) on a 5-point scale. For example, a score of ‘4’ indicates a very high impact, a score of three indicates a very high impact, a score of two indicates a low impact, a score of one indicates a very low impact, and a score of “0” indicates no impact. Experts establish the connections between the diverse variables.

$$M = \begin{bmatrix} M & \cdots & M_{n1} \\ \vdots & \ddots & \vdots \\ M_{1n} & \cdots & M \end{bmatrix}$$

Table 2. Direct relationship matrix of Respondent 1.

	ES1	ES2	ES3	ES4	ES5	ES6	ES7	ES8
ES1	0	1	0	1	2	1	1	2
ES2	4	0	1	1	3	3	4	4
ES3	4	3	0	1	1	3	1	3
ES4	1	1	4	0	3	2	3	4
ES5	4	3	4	2	0	3	0	2
ES6	1	2	2	2	0	0	4	3
ES7	0	2	1	1	1	2	0	2
ES8	0	1	0	2	1	3	1	0

Source: Author’s Own

Step II Average relation matrix (AV) Calculation: Equation 7.1 is used to calculate the average relation matrix (Table 3) based on the direct relationship matrices provided by the five respondents.

$$AV = (M1 + M2 + \ldots\ldots\ldots\ldots + M5)/5 \qquad (7.1)$$

Table 3. Average relationship matrix.

Average	ES1	ES2	ES3	ES4	ES5	ES6	ES7	ES8
ES1	0	1	0	1	1.6	1	1	2
ES2	3.6	0	1	0.6	3	2.6	3.4	3.8
ES3	4	3	0	1.2	1.8	2.8	1	3.2
ES4	1	1	3	0	3	2	3	4
ES5	3.8	2.8	3.2	2.2	0	3	0	2.2
ES6	1	2.2	2	1.8	0	0	3.8	2.8
ES7	0	2	1	1	1	2	0	2
ES8	0	0.8	0	2	1	2.6	1	0

Source: Author's Own

Step III: Compute the normalized matrix (N): In the third step, the average relation matrix (AV) (provided in Eq. 7.3) is multiplied by a constant 'k' (derived from equation 7.2) to calculate the normalized matrix (N) (Raghuvanshi et al. 2017, Chakrabotry et al. 2018, Raghuvanshi and Agrawal 2020, Ranjan et al. 2016) (Table 4):

$$k = Minimum\left[\frac{1}{Max\sum_{j=1}^{n} AV_{ij}}, \frac{1}{Max\sum_{i=1}^{n} AV_{ij}}\right] \tag{7.2}$$

$$N = k * \text{AV} \tag{7.3}$$

Constant is represented by "k",

AV_{ij} = average relationship matrix element.

"i"= element in row

"j"= element in column

Table 4. Normalized matrix.

N	ES1	ES2	ES3	ES4	ES5	ES6	ES7	ES8
ES1	0	0.05	0	0.05	0.08	0.05	0.05	0.1
ES2	0.18	0	0.05	0.03	0.15	0.13	0.17	0.19
ES3	0.2	0.15	0	0.06	0.09	0.14	0.05	0.16
ES4	0.05	0.05	0.15	0	0.15	0.1	0.15	0.2
ES5	0.19	0.14	0.16	0.11	0	0.15	0	0.11
ES6	0.05	0.11	0.1	0.09	0	0	0.19	0.14
ES7	0	0.1	0.05	0.05	0.05	0.1	0	0.1
ES8	0	0.04	0	0.1	0.05	0.13	0.05	0

Source: Author's Own

Step IV: Total-relation matrix (T) Calculation: Equation (7.4) is applied to derive the total-relation matrix from the normalized matrix (Table 5).

$$T = N(I - N)^{-1} \tag{7.4}$$

Where, I is an identity matrix.

Table 5. Total relationship matrix.

T	ES1	ES2	ES3	ES4	ES5	ES6	ES7	ES8
ES1	0.08042	0.13198	0.07158	0.12206	0.14904	0.15851	0.13954	0.22152
ES2	0.33216	0.18485	0.18582	0.18786	0.28845	0.35043	0.34413	0.44084
ES3	0.35685	0.31234	0.13314	0.20780	0.24182	0.35324	0.24682	0.41863
ES4	0.23040	0.24087	0.28284	0.16026	0.29331	0.33470	0.32714	0.45795
ES5	0.37026	0.31854	0.29106	0.25930	0.17335	0.37629	0.21758	0.39949
ES6	0.17906	0.24588	0.19949	0.20180	0.13372	0.18820	0.33432	0.35139
ES7	0.10735	0.19960	0.13217	0.13729	0.14003	0.23030	0.12312	0.25722
ES8	0.08349	0.12935	0.08281	0.16960	0.12392	0.23228	0.15698	0.14194

Source: Author's Own

Step V: Cause-and-effect relationship table: Equations (7.5) and (7.6) are utilized to construct the cause-and-effect relationship table, as presented in Table 6.

R represents the sum of rows and
C represents the sum of columns.

$$R = \left[\sum_{j=1}^{n} T_{ij}\right]_{n\times 1} \tag{7.5}$$

$$C = \left[\sum_{l=1}^{n} T_{ij}\right]_{1\times n} \tag{7.6}$$

Table 6. Cause-and-effect relationship

	R	C	R+C	R-C
ES1	1.074655	1.739987	2.814642	–0.66533
ES2	2.314552	1.763405	4.077957	0.551148
ES3	2.270646	1.378926	3.649572	0.891719
ES4	2.327462	1.445974	3.773436	0.881487
ES5	2.405868	1.54364	3.949507	0.862228
ES6	1.833867	2.223966	4.057833	–0.3901
ES7	1.327079	1.889623	3.216702	–0.56254
ES8	1.120378	2.688986	3.809365	–1.56861

Source: Author's Own

Step VI: Draw the causal diagraph: A cause-and-effect table is employed to generate a causal diagraph. The horizontal and vertical axes are used to calculate the R+C and R–C values, C denotes the overall effect of element i on element j, while R represents the overall influence of element i on element j. The criteria are categorized as effect if the (R–C) value is negative, and as cause when the (R–C) value is positive.

4. Results and Analysis

The present study employed the multi-criteria decision-making technique, DEMATEL, to determine the ranking and causal relationships among the barriers to energy security (ES). A threshold value of 0.32 was selected to filter out insignificant interactions among the factors. This threshold was established by summing the mean and standard deviation of the total relationship matrix, thereby eliminating trivial influences. The following barriers of ES were prioritized based on R + C values: ES2 > ES6 > ES5 > ES8 > ES4 > ES3 > ES7 > ES1.

R-C values are used to prioritize causal barriers, which are as follows: ES3 > ES4 > ES5 > ES2. Limited Knowledge of energy conservation and efficiency (ES3) is the most noticeable barrier to energy security which led to community resistance (ES1), undervaluing traditional knowledge integration (ES6), and insufficient expertise in energy efficient programs (ES8). Second important barrier to energy security has been identified as Low financial and energy awareness (ES4), which results in the undervaluing traditional knowledge integration (ES6), limited project development experience (ES7), and insufficient expertise in energy efficient programs (ES8). The third important barrier to energy security is identified as limited decision making capability (ES5) which led to the community resistance (ES1), undervaluing traditional knowledge integration (ES6), and insufficient expertise in energy efficient programs (ES8). And the fourth important barrier to energy security is identified as insufficient community involvement (ES2) which led to all the resultant barriers the community resistance (ES1), undervaluing traditional knowledge integration (ES6), limited project development experience (ES7), and insufficient expertise in energy efficient programs (ES8). The rest of the barriers are the effect barriers or we can say the resultant of the causal barriers: Undervaluing the traditional knowledge integration (ES6); limited project development experience (ES7), community resistance (ES1) and insufficient expertise in energy efficient programs (ES8).

Undervaluing the traditional knowledge integration (ES6) and insufficient expertise in energy efficient programs (ES8) is discovered as outcome to all the four causal barriers, followed by community resistance (ES1), and limited project development experience (ES7).

Figure 1 shows the cause-and-effect relationship between causal barrier and their resultant barriers. To be more precise, each effect causal barrier leads to which specific resultant barrier has been shown in Figure 1.

The causal barriers located above the x-axis exert the most influence on the resultant barriers situated below the x-axis, referred to as influenced, resultant, or causative factors, as illustrated in Figure 1. Moreover, all elements have been categorized into four distinct zones (Figure 2) for a more accurate analysis of causal

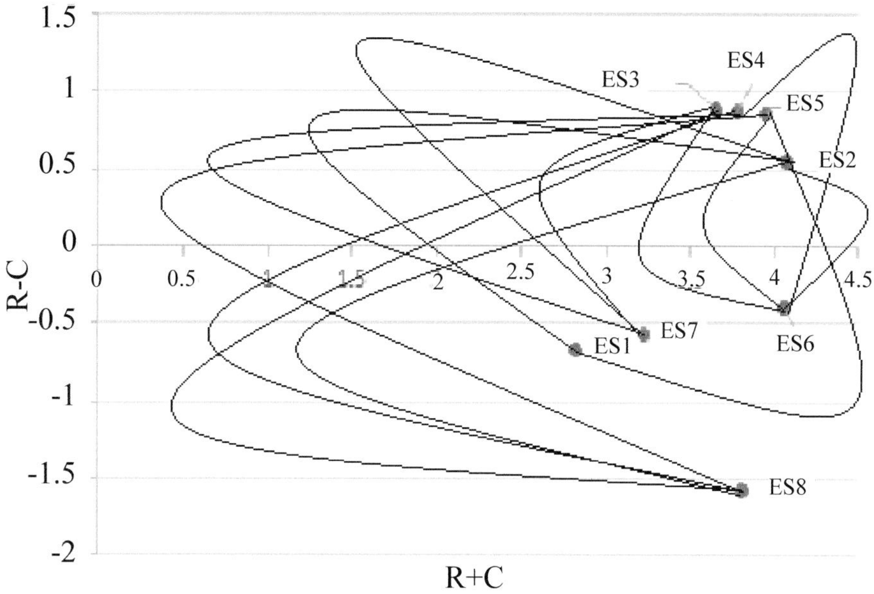

Figure 1. Cause and Effect diagraph of the barriers of energy security. Source: Author's Own.

and consequent factors (zones 1, 2, 3, and 4). Zone 1 signifies the outcome barriers that exert the minimal influence compared to all other barriers. The barriers in zone 1 exhibit an unstable interaction with other causal barriers. None of the obstacles belong to this category. Zone 2 signifies barriers that exert a moderate influence on future barriers; no factors are classified under this area.

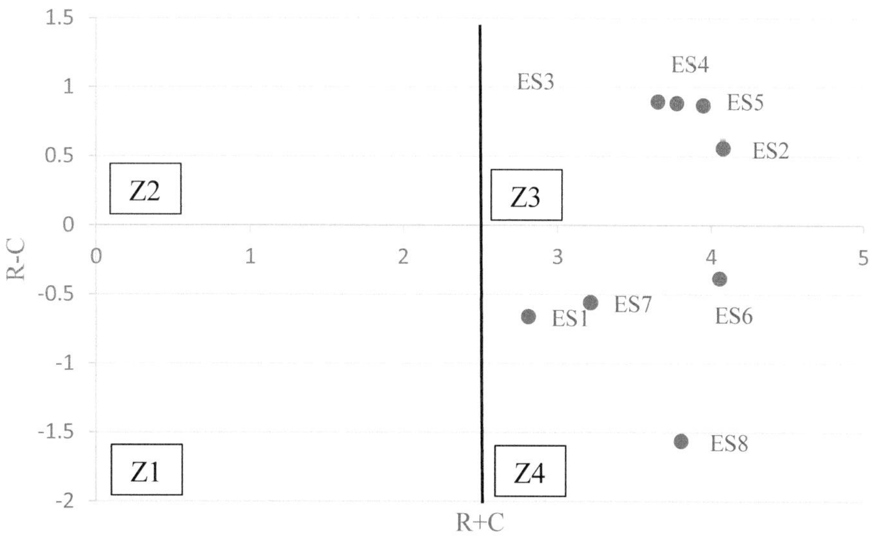

Figure 2. Zone wise representation of energy security barriers. Source: Author's Own.

Zone 3 comprises the most influential barriers. These barriers are closely linked to the outcomes and are considered the most critical causal factors. All the causal barriers positioned in this particular zone. This zone includes limited Knowledge of energy conservation and efficiency (ES3), low financial and energy awareness (ES4), limited decision making capability (ES5), insufficient community involvement (ES2).

Zone 4 is made up of subsequent barriers that have a strong influence of the cause factor in zones 2 and 3. Undervaluing the traditional knowledge integration (ES6); limited project development experience (ES7), community resistance (ES1) and insufficient expertise in energy efficient programs (ES8) are the resultant barriers placed under zone 4.

5. Discussion and Implications

Limited Knowledge of Energy Conservation and Efficiency has been identified as the most crucial barrier among all the identified barriers. Due to lack of knowledge around conservation and efficient energy use directs to unsustainable practices, which strain energy resources and decrease overall system reliability. Communities can better manage energy consumption by promoting awareness of efficient practices, by helping to ensure a more ways for sustainable energy supply.

The second most important limitations which require the immediate attention/ focus is the low financial and energy awareness. For better understanding of costs and benefits of energy solutions, financial and energy literacy lay the foundation, influences the community support and involvement. Also, resistance in fostering of modern energy services might be due to lack of awareness which plays a vital role in ineffective resource management, therefore hindering/lacking energy security efforts.

Then, the third casual barriers of the energy security has been emphasized as limited decision-Making Capability. This aspect is the approachable one, as limited decision-making capability among local leaders or community members avoid the initiation and advancement of energy projects. Communities are unable to establish goals, allocate resources, or commit to new energy solutions due to lack or absent of empowered decision-makers, which directly results in impacting their ability to support or adopt energy initiatives.

Despite all, the insufficient community involvement has been seen depicting as a another causal and approachable barriers to the several other barriers. Projects risk are being poorly aligned with local needs and may lack community support as due to lack of active community engagement, which plays an important role for success. This aspect depicts as a root cause by creating a isolation between the community and project developers, as it leads or results in impacting the project's conformity and long-term sustainability.

5.1 Theoretical Implications

This chapter emphasizes the theoretical understanding and significance of energy security by recognizing and prioritizing the crucial barriers influencing sustainable energy practices. The identified barriers such as limited knowledge of energy conservation and efficiency, low financial and energy awareness, limited decision-making capability, and insufficient community involvement provides a comprehensive prospectus of the causal linkages between these factors.

This research highlight and emphasizes the importance of understanding and consciousness for improving energy security as it focuses on the theoretical framework unlike the previous studies by conjoining the findings from the prior research, along with the significance of decision-making and community engagement. On the other hand, it delineates causal relationships among barriers and emphasises their comparative impact by prioritising the barriers through a DEMATEL methodology. This methodology distinguishes and impacts the study by offering a systematic and pragmatic framework for handling energy security issues. Therefore, this work increases the academic discourse on energy security by combining its findings with existing literature, establishing a foundational reference for further research in comparable settings.

5.2 Practical Implications

This chapter provides a significant understanding by building knowledge and awareness for policy makers, energy project developers, and community leaders. Initiatives like educational initiatives which should be first focused on to enhance comprehension of energy conservation and efficiency practices.

It can enable communities to make informed choices regarding contemporary energy solutions by developing financial and energy literacy, and also encourages more adoption and support to it. Secondly, by enhancing decision-making capabilities it equips local leaders and community people with training and resources to get better and improve their proficiency in building and managing energy projects. Because of this, the creation of collaborative and cooperating efforts the platforms that facilitate transparent and inclusive decision-making among stakeholders is equally plays an important role.

For especially in developing nations, this paper identified barriers and attaining sustainable energy security by outlining a strategy for surrounding.

6. Conclusion

The present study identified 8 barriers to energy security based on the existing literature. Utilizing the DEMATEL approach, these factors were ranked and further the structure of their causal linkages were determined. Though previously researchers attempted to discover the barriers of energy security but in fragments, however, the current study was aimed at establishing a comprehensive framework and generating a deeper understanding by prioritizing the causal link structures of energy security barriers. Precisely, the current study contends that limited Knowledge of energy conservation and efficiency, low financial and energy awareness, limited decision making capability, insufficient community involvement emerged as major causal barriers to energy security and leading to their resultant barriers such as Undervaluing the traditional knowledge integration, limited project development experience, community resistance and insufficient expertise in energy efficient programs.

7. Limitation and Future Scope

When interpreting the findings of this study, several limitations should be considered. Initially, the causal structure framework was constructed by analyzing only eight energy security barriers from the literature. It will be imperative to expand this list in future research in order to refine and develop the present causal linkage structure of energy security. Secondly, the generalizability of these findings may be improved by incorporating a larger sample size in future research, as this study was reliant on the responses of five experts. Furthermore, although DEMATEL was implemented in this investigation, future investigations may integrate sophisticated statistical methodologies and other multi-criteria decision-making methodologies, including AHP, grey-DEMATEL, T-ISM, and TOPSIS. Lastly, future research may incorporate a more comprehensive inventory of energy security barriers and respondents from a variety of industries and countries to enable comparative studies.

References

Ahlborg, H. and Hammar, L. 2014. Drivers and barriers to rural electrification in Tanzania and Mozambique–Grid-extension, off-grid, and renewable energy technologies. Renewable Energy 61: 117–124.

Armaroli, N. and Balzani, V. 2007. The future of energy supply: challenges and opportunities. Angewandte Chemie International Edition 46(1-2): 52–66.

Barnett, A. 2000. Energy and the Fight Against Poverty. Department for International Development (Dfid), Livelihood Sector Report, UK.

Berkes, F. 2012. Sacred Ecology (3rd ed.). New York: Routledge.

Bhattacharyya, S. C. and Palit, D. (eds.). 2014. Mini-grids for Rural Electrification of Developing Countries: Analysis and Case Studies from South Asia. Springer.

Chakraborty, S., Chatterjee, P. and Prasad, K. 2018. An integrated DEMATEL–VIKOR method-based approach for cotton fibre selection and evaluation. Journal of the Institution of Engineers (India): Series E 99: 63–73.

Chaurey, A. and Kandpal, T. C. 2010. Assessment and evaluation of PV based decentralized rural electrification: An overview. Renewable and Sustainable Energy Reviews 14(8): 2266–2278.

Chaurey, A., Krithika, P. R., Palit, D., Rakesh, S. and Sovacool, B. K. 2012. New partnerships and business models for facilitating energy access. Energy Policy 47: 48–55.

Fukutomi, M. 2024. Oil or geopolitical issues?. Quantitative rethinking of political instability in the Middle East and North Africa. GeoJournal 89(2): 55.

Hancock, K. J. and Vivoda, V. 2014. International political economy: a field born of the OPEC crisis returns to its energy roots. Energy Research & Social Science 1: 206–216.

Ibekwe, K. I., Etukudoh, E. A., Nwokediegwu, Z. Q. S., Umoh, A. A., Adefemi, A. and Ilojianya, V. I. 2024. Energy security in the global context: A comprehensive review of geopolitical dynamics and policies. Engineering Science & Technology Journal 5(1): 152–168.

International Energy Agency (IEA). 2007a. Energy Security and Climate Policy: Assessing Interactions. Paris: OECD/IEA.

International Energy Agency (IEA). 2007b. Oil Supply Security: Emergency Response of IEA Countries. Paris: OECD/IEA.

International Energy Agency (IEA). 2007. World Energy Outlook 2007. Paris: OECD/IEA.

Janda, K. B. and Parag, Y. 2013. A middle-out approach to understanding the role of individuals in energy demand: A case study of energy efficiency in the UK. Energy Research & Social Science 6: 36–43.

Lee, L. S., Lee, Y. F., Wu, M. J. and Pan, Y. J. 2019. A study of energy literacy among nursing students to examine implications on energy conservation efforts in Taiwan. Energy Policy 135: 111005.

Lubell, H. 1961. Security of supply and energy policy in Western Europe. World Politics 13(3): 400–422.

Luthra, S., Kumar, S., Garg, D. and Haleem, A. 2015. Barriers to renewable/sustainable energy technologies adoption: Indian perspective. Renewable and Sustainable Energy Reviews 41: 762–776.
Mandelli, S., Barbieri, J., Mattarolo, L. and Colombo, E. 2016. Sustainable energy in Africa: A comprehensive data and policies review. Renewable and Sustainable Energy Reviews 58: 1621–1646.
Matinga, M. N. and Annegarn, H. J. 2013. Paradoxical impacts of electricity on life in a rural South African village. Energy Policy 58: 295–302.
Modi, V., McDade, S., Lallement, D. and Saghir, J. 2005. Energy services for the Millennium Development Goals. The International Bank for Reconstruction and Development/The World Bank and the United Nations Development Programme. Available from: https://documents1.worldbank.org/curated/en/574701468321304694/pdf/346500PAPER0Energy0services01PUBLIC1.pdf.
Pachauri, S. and Spreng, D. 2012. Towards an integrative framework for energy transitions of households in developing countries. Tackling Long-Term Global Energy Problems: The Contribution of Social Science, 73–96.
Qazi, A., Bhowmik, C., Hussain, F., Yang, S., Naseem, U., Adebayo, A. A. et al. 2021. Analyzing the public opinion as a guide for renewable-energy status in Malaysia: A case study. IEEE Transactions on Engineering Management 70(2): 371–385.
Raghuvanshi, J. and Agrawal, R. 2020. Revitalization of Indian SMEs for sustainable development through innovation. Business Strategy & Development 3(4): 461–473.
Raghuvanshi, J., Agrawal, R. and Ghosh, P. K. 2017. Analysis of barriers to women entrepreneurship: The DEMATEL approach. The Journal of Entrepreneurship 26(2): 220–238.
Ramachandra, T. V. 2009. RIEP: Regional integrated energy plan. Renewable and Sustainable Energy Reviews 13(2): 285–317.
Ranjan, R., Chatterjee, P. and Chakraborty, S. 2016. Performance evaluation of Indian Railway zones using DEMATEL and VIKOR methods. Benchmarking: An International Journal 23(1): 78–95.
Reddy, B. S. and Balachandra, P. 2006. Dynamics of technology shifts in the household sector—implications for clean development mechanism. Energy Policy 34(16): 2586–2599.
Reed, M. S. 2008. Stakeholder participation for environmental management: A literature review. Biological Conservation 141(10): 2417–2431.
Singh, S. 2021. Energy crisis and climate change: Global concerns and their solutions. Energy: Crises, Challenges and Solutions, 1–17.
Smil, V. 2007. Energy in Nature and Society: General Energetics of Complex Systems. MIT Press.
Sovacool, B. K. 2012. The political economy of energy poverty: A review of key challenges. Energy for Sustainable Development 16(3): 272–282.
Sovacool, B. K. and Dworkin, M. H. 2015. Energy justice: Conceptual insights and practical applications. Applied Energy 142: 435–444. https://doi.org/10.1016/j.apenergy.2015.01.002.
United Nations Environment Programme (UNEP). 2011. Towards a Green Economy: Pathways to Sustainable Development and Poverty Eradication. Nairobi: United Nations Environment Programme. Available from. https://www.unep.org/resources/report/towards-green-economy-pathways-sustainable-development-and-poverty-eradication.
Van der Linde, C., Amineh, M. P., Correljé, A., de Jong, D. and Hansenr, S. 2004. Study on energy supply security and geopolitics. Available at https://policycommons.net/artifacts/11281431/study-on-energy-supply-security-and-geopolitics/12166663/.
Van Veelen, B. 2018. Negotiating energy democracy in practice: governance processes in community energy projects. Environmental Politics 27(4): 644–665.
Wlokas, H. L., Boyd, A. and Andolfi, M. 2012. Challenges for local community development in private sector-led renewable energy projects in South Africa: an evolving approach. Journal of Energy in Southern Africa 23(4): 46–51.
Yergin, D. 2006. Ensuring energy security. Foreign Affairs 69–82.
Zhou, Q., Huang, W. and Zhang, Y. 2011. Identifying critical success factors in emergency management using a fuzzy DEMATEL method. Safety Science 49(2): 243–252.

Chapter 8

Using Building Information for Building Energy Modelling

Vinayak Vishwakarma

1. Introduction to BIM and BEM

1.1 Context and Evolution of BIM

Building Information Modelling has become a major technological improvement in Architecture, Engineering, and Construction work that helps designers and builders visualise projects better while making data easier to use. Building Information Modelling started in 2000 as a major shift from traditional design processes thanks to its advanced ability to handle project data through all building stages, including planning, construction, maintenance, and updates. BIM models handle more project data than past systems and let teams get precise updates about material stock, construction schedules, and expenses for better teamwork. Large-scale building projects need BIM tools to run better and make workflow improvements (Ambrosini et al. 2023).

Building Information Modeling (BIM) encountered multiple major obstacles during its initial rollout phase, which impeded its broad adoption by stakeholders. The adoption of BIM encountered three major obstacles caused by expensive set-up costs, low understanding of BIM advantages, and undertrained stakeholders. Many organisations declined to invest in BIM because they perceived high uncertainty about the return on investment (ROI), according to Ismail et al. (2017). A cultural resistance towards new methods persisted in the construction industry despite the move towards BIM practices because traditional methods were preferred. Interoperability challenges between various software systems and the lack of standardised tools both

Associate Professor – Operations, Jaipuria Institute of Management, Indore, India.
Email: vinayakvishwakarma@gmail.com

DOI: 10.1201/9781003475385-8

contributed to the slow adoption rate, according to McAuley et al. (2017). The UK, together with other regions, demonstrated improved BIM adoption when public awareness combined with specific official regulations began to take effect.

1.2 Transition from CAD to BIM and its Implications

BIM signifies a substantial advancement from conventional Computer-Aided Design (CAD), which principally emphasises the generation of 2D or 3D design drawings devoid of real-time data integration. Although CAD transformed the AEC sector by digitising blueprints and improving design precision, it lacked multi-dimensional data storage and communication capabilities. BIM facilitates a holistic, multi-dimensional model that incorporates time (4D), cost (5D), and sustainability considerations (6D). This transition enables architects, engineers, and contractors to collaborate on a unified model, utilising real-time data for improved decision-making. The adoption of BIM has facilitated efficient project processes, minimised errors, and enhanced risk management, establishing it as a vital instrument in contemporary construction (Fonseca Arenas and Shafique 2023)

1.3 Necessity of Integrating BIM with BEM for Sustainable Design

As people focus more on sustainable practices, BIM and Building Energy Modelling must work together to help designers create energy-efficient buildings that support the environment. BEM models help us measure how well a building uses energy by examining its heating and cooling systems, plus lighting systems. Separate BEM applications create design mismatches when they are not part of the full design process. When BEM works with BIM technology teams, they can test energy-saving strategies before the design stage ends. Under environmental regulations like the European Union's Green Homes Directive, firms must unite digital building methods to meet new energy standards and cut building expenses (Skrzypczak et al. 2023, Daniels 2023).

2. Limitations of Traditional Building Energy Modelling (BEM)

The conventional Building Energy Modelling system produces valuable building performance information. Building Energy Modelling shows its growing problems because current sustainability needs are higher. Traditional building energy modelling usually does not work well with the early architectural design phases because the two processes remain separate. Building Energy Modelling (BEM) reviews become available too late to influence the design because the evaluation happens after the design work is complete. This approach produces suboptimal results because Building Energy Modelling does not include the necessary feedback loop to direct sustainable design decisions during the early design stages (Cheng et al. 2024, Fonseca Arenas and Shafique 2023).

Standard BEM needs a lot of data input and manual handling, which makes it harder to work with. Each design project requires separate steps to select architectural forms, materials, and system components. The manual energy model input process is long and often produces modelling errors due to design adjustments or human mistakes (Eames and Vinai 2024). The absence of real-time updates makes the model less precise, so it cannot show current design changes, which limits the model's value in achieving energy efficiency (Skrzypczak et al. 2023). Conventional BEM struggles because its software platforms fail to work together. Teams need separate business software that works poorly together, creating data silos. BIM allows teams to share and update data simultaneously.

In contrast, traditional BEM operates in separate environments and blocks collaboration between architects, engineers, and contractors. BEM cannot help designers use energy efficiently because of this issue. Standalone BEM tools prevent industry progress towards combined work processes that meet energy compliance standards effectively (Ambrosini et al. 2023, Daniels 2023). The European Union's Green Homes Directive pushes sustainability standards beyond previous limits, so BIM and BEM must work together. The data connection between BIM and BEM during early project stages makes energy insights more accurate and improves design quality to help achieve sustainable building performance (Cheng et al. 2024, Fonseca Arenas and Shafique 2023, Eames and Vinai 2024).

The evaluation of project feasibility demands a clear comprehension of BIM high-performance simulation cost components. Implementing high-performance simulations demands investments in specialised software and hardware systems together with prolonged expenses on employee training for optimised tool operation. High-performance simulations result in substantial costs that extend both project schedules and financial budgets. High-performance simulations bring improved design decisions and energy efficiency, but they lengthen modelling phases, which ultimately stretch project timelines. The evaluation of financial implications for BIM stakeholders requires an in-depth analysis of costs involved with high-performance simulation integration.

Through different steps, BIM and BEM work together to connect design teams with simulation and energy analysts for better stakeholder collaboration. The standard BIM-BEM process starts with design when architects and engineers create a digital model that contains basic building dimensions, plus material and structural information. The initial BEM team model allows for detailed energy simulations by analysing heating, cooling, and lighting needs (Refer Figure 1).

Our energy modelling work begins right after creating the design. At this time, BEM tools show how energy use behaves and help designers see which materials and systems work best. During analysis, simulations reveal where to enhance insulation performance and adjust HVAC systems for better results. The team keeps repeating this process until it finds an energy-efficient design that follows regulation requirements. At project handover, facility managers receive the final BIM-BEM model with all approved data to help them maintain building energy performance throughout operations (Eames and Vinai 2024).

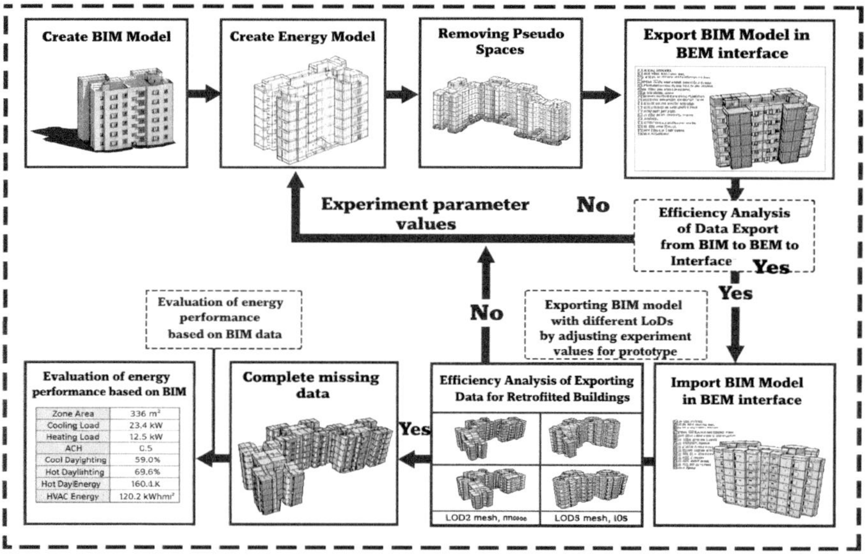

Figure 1. BIM-BEM integration process framework (Abd-Elrahman H. A. Abd-Elnaby and Rabee M. Reffat 2024).

3. Data Interoperability

Data interoperability is an essential component of BIM-BEM integration because it ensures that data flows smoothly between systems (Refer Figure 2). The BIM-BEM process depends on standard data-transformation methods that let different platforms understand each other. The standard data exchange formats IFC and gbXML enable BIM data to work with BEM tools through their translation process. Research demonstrates that converting data presents difficulties since material characteristics and specific HVAC parameter information can get damaged or disappear during the process (Ambrosini et al. 2023).

Middleware products help BIM-BEM integration work by connecting different software systems that cannot talk to each other directly. These tools help link BIM and BEM systems by transforming BIM data into easy-to-process files that BEM programs accept to produce a single energy modelling process. The customisation required for middleware solutions when fitting into project specifications adds time and expense to project workflow (Skrzypczak et al. 2023). The integration of Building Information Modelling (BIM) with Building Energy Modelling (BEM) now utilises middleware solutions to solve interoperability problems. Different software applications connect through middleware, which establishes efficient two-way communication between platforms. Different system compatibility barriers disable effective collaboration between stakeholders; thus, the implementation of various interoperability solutions becomes necessary. Organisations achieve better BIM-BEM integration through the use of middleware solutions, which both improve data coherence and simplify workflow processes. Detailed insights into particular middleware technologies, together with their implementation examples,

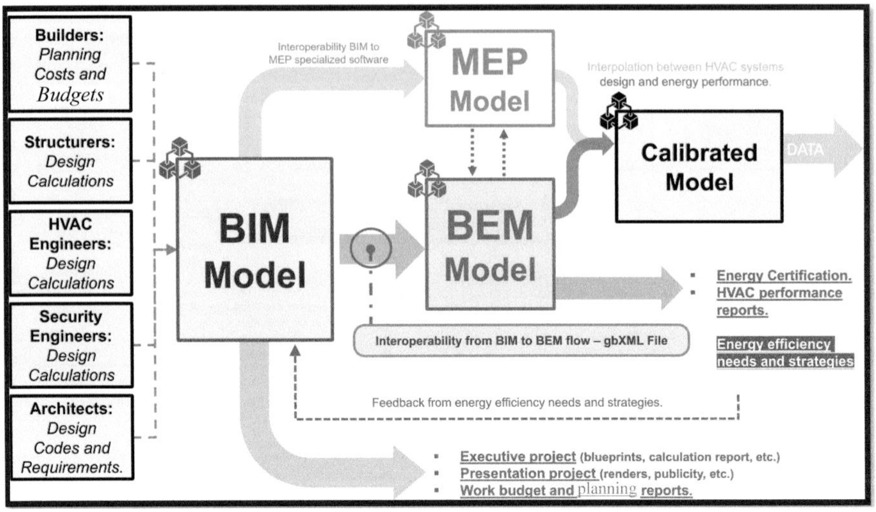

Figure 2. BIM data workflow scheme for other platforms (Adopted from Guízar Dena et al. 2024).

will demonstrate how these solutions resolve interoperability issues in construction sector processes.

3.1 The Interchange of Data Through IFC and gbXML Standards

The proper functioning of BIM-BEM tools depends on interoperability standards such as IFC and gbXML. buildingSMART ensures the long-lasting vitality of IFC as an open standard that enables BIM applications to work together with full building data. The gbXML standard was developed to change BIM information into formats that BEM tools can use while assisting in energy analysis. New research reveals performance issues with these established standards. The ability of both IFC and gbXML to transfer basic building information remains strong, but their accuracy in transmitting comprehensive energy details remains limited according to Cheng et al. (2024) and Eames and Vinai (2024).

Even though standards keep BIM and BEM working together, they need regular updates to prevent energy model data from getting lost and to show building parameters correctly. As global energy and environmental rules get more complex, these standards need further improvement to work properly.

4. BIM-BEM Integration Applications and Project Examples

The combination of BIM and BEM tools now lets designers and constructors build better energy-efficient buildings. The combined approach allows design teams to test building performance while making better design choices to cut down energy usage.

4.1 Case Studies on Energy-Efficient Design Utilizing BIM-BEM

4.1.1 Energy-Efficient Ski Lodges

Snow sports resorts in frigid areas need to use extra energy to heat their buildings. Through BIM-BEM integration, the architects designed lodges that use modern insulation materials while positioning rooms for optimal sunlight and including solar power technology. Through BIM-BEM integration, a ski lodge in Norway achieved 30% less heating energy use and saved €50,000 in annual energy expenses. Our BEM simulations showed how to improve envelope materials, which lessened heat loss by 20%.

4.1.2 Commercial Office Buildings

Business buildings gain major advantages from BIM and BEM integration by optimizing HVAC systems plus lighting and insulation choices. Singapore's commercial office building used BIM-BEM tools to study lighting performance and then installed real-time monitoring technology. By using BIM-BEM integration, the project saved 25% in energy costs for running operations. By using BEM to enhance thermal modelling, the building achieved better air quality and met regional green requirements.

4.1.3 Residential Structures

BIM-BEM integration helps designers in residential projects make better choices about insulation materials and HVAC systems while finding the best window locations. The German residential complex analysis showed how energy modelling picked the right combination of wall size and window location to decrease yearly heating and cooling power usage by 18%. Our life cycle analysis showed that the project would need 15% less maintenance during the next 20 years.

4.1.4 Real-Time Monitoring and Control

With BIM-integrated Internet of Things services, building systems get real-time status updates that allow better energy management through BEM. In a UK pilot project, digital twin monitoring helped reduce HVAC energy usage by 20% by tracking system performance in real time. Predictive models helped maintain equipment better, which increased lifetime operation by 10%. Life Cycle Cost Analysis helps companies see the full expenses of all building phases. Life Cycle Cost Analysis combines BIM and BEM data to show all construction and operations costs plus planned upgrades. A U.S. university campus project earned LEED certification when BIM-BEM integration revealed 35% less energy use over 10 years to meet sustainability targets and save money.

Building-specific examples that demonstrate IoT and Digital Twin implementations for lifecycle energy management effectively show the operational advantages of these systems. The Edge in Amsterdam demonstrates IoT infrastructure excellence by using real-time monitoring to track energy consumption and environmental performance, positioning it as one of the world's most environmentally friendly buildings. Real-time data collection from the building allows its management system to perform dynamic energy optimisation, delivering a reported 70% reduction in energy costs compared to standard buildings. Digital Twin technology in New York City's Hudson Yards development enables developers to construct digital representations of physical assets that drive predictive maintenance solutions for improved energy performance. The expense analysis of deploying these advanced systems should be studied, especially for medium and small buildings that struggle to invest in robust technological frameworks for management. The evaluation of IoT and Digital Twin investment benefits and operational cost reductions allows stakeholders to make informed choices regarding the adoption of innovative solutions.

By combining BIM and BEM in sustainable building design, teams get all the data needed to meet LEED and BREEAM certification standards while maintaining energy performance measurements, water consumption reporting, and materials tracking. Using BIM and BEM together creates better buildings while producing fewer carbon emissions, helping to combat climate change. Joint BIM and BEM use leads to valuable ways to design buildings that save energy and operate sustainably. Through the joint use of these technologies, designers can build green buildings that work well for the environment and save money. As the building industry evolves, BIM+BEM integration will become vital to fighting climate change and reducing energy use.

5. Challenges in BIM-BEM Integration

This section builds on the established knowledge base to study BIM and BEM integration challenges and technical solutions. It examines how BIM and BEM systems face technical problems and presents advanced technology solutions.

The technical system and everyday work operations create problems for people.

- *Data Compatibility and Interoperability*: AEC professionals use different BIM and BEM programs that work with specific data formats, such as Revit and ArchiCAD for BIM, plus EnergyPlus and eQuest for BEM. These systems have persistent problems exchanging information, creating repeated data compatibility issues that slow down work and make data less reliable (Ambrosini et al. 2023, Skrzypczak et al. 2023).
- *Absence of Standardized Data formats*: The lack of standard data formats for energy components in building models makes energy modelling difficult to achieve smoothly. Though IFC and gbXML standards strive to fix this problem, they still struggle to deliver precise thermal and performance data results because of design restrictions (Cheng et al. 2024).
- *Data Quality and Consistency*: BIM-BEM projects depend on consistent, high-quality data for accurate energy calculation results. Poor data quality may create inaccurate energy efficiency results that violate increasingly strict energy rules like the EU's Green Deal (Fonseca Arenas and Shafique 2023).

- *Improved Building Envelope Design*: Through BIM and BEM integration, designers can check their building envelope performance by testing insulation materials, window systems, and air leakage sealing features. Designers can enhance both energy savings and system function by altering building parts.
- *Real-Time Surveillance and Regulation*: Our ability to control building systems in real-time improves when we link BIM to building automation systems. This system finds problems before they become severe and helps us save energy while making buildings more comfortable.
- *Deficiency of Proficient Experts*: The convergence of BIM and BEM necessitates expertise in both digital modelling and energy analysis, a competency that is still limited. The shortfall results in project timetable delays and may jeopardise the validity of energy modelling outcomes (Ambrosini et al. 2023).
- *Training and Educational Requirements*: Industry training programmes for integrating BIM and BEM are scarce. Extensive, practical training in both BIM and BEM technologies might greatly enhance the workforce capable of managing these responsibilities; nevertheless, the existing availability is limited, highlighting the necessity for industry investment in training (Cheng et al. 2024, Fonseca Arenas and Shafique 2023).

6. Emerging Trends and Future of BIM-BEM

BIM-BEM merges with machine learning technology to improve performance. Artificial intelligence and machine learning now help BIM-BEM integration by improving energy simulations and modelling better building designs. The main use of AI technology lies in its ability to run energy simulation tasks automatically. The standard method of energy modelling demands repeated complex computational work to explore multiple power usage setups. AI tools can run simulation operations much faster than humans do. With AI algorithms running in real-time, they scan designs for energy performance weaknesses, which allows prompt material and design modifications (Fonseca Arenas and Shafique 2023).

Machine learning helps BIM-BEM predict how much energy buildings will use. Machine learning models create energy usage forecasts by studying past building energy data combined with occupancy patterns, environmental conditions, and design parameters. With this predictive power, designers can identify peak energy use times to add efficiency features ahead of time. Advanced AI systems analyse design layouts to suggest effective energy-saving recommendations for building spaces and systems, including HVAC systems, which help architects create greener structures (Ambrosini et al. 2023, Skrzypczak et al. 2023).

AI tools help teams process project data better to enhance BIM-BEM integration. By finding energy patterns, AI helps to make real-time data-based choices and suggests the best energy solutions. These tools help connect BIM-BEM by decreasing manual data tasks and simplifying project processes, especially for large projects. Reinforcement learning algorithms now help AI tools refine building energy models more accurately by adjusting them to handle complex systems (Eames and Vinai 2024).

Artificial Neural Networks (ANN), Support Vector Machines (SVM), and Decision Trees (DT) represent the main supervised learning models that experts use to forecast energy consumption based on historical data. The training of these models on labelled datasets enables them to detect patterns that allow predictions about upcoming energy consumption levels. Research demonstrates that ANN delivers superior results over alternative algorithms when used for annual energy consumption forecasts because it reaches higher R-squared metrics, indicating data variability explanation (Amasyali and El-Gohary 2017).

Unsupervised learning approaches detect energy usage patterns by analysing data sets that lack predefined classification labels. The techniques discover latent patterns in data to help researchers better understand the energy consumption patterns of buildings under varying circumstances.

Reinforcement learning represents an emerging method that allows algorithms to discover optimal energy management strategies by executing repeated trials. Building operations become more efficient over time by implementing this method, which reacts to real-time energy system input.

The implementation of machine learning methodologies within BIM-BEM integration produces substantial improvements in energy forecasting functionality. Building design sustainability receives a boost from these predictive models, which help designers forecast energy peak moments so they can introduce proactive efficiency measures (Ambrosini et al. 2023, Skrzypczak et al. 2023). The investigation of machine learning models throughout this paper answers the reviewer while demonstrating their essential function in energy optimisation for contemporary architectural designs.

6.1 The Role of Digital Twins and IoT in Future BIM-BEM Developments

Digital twins and the Internet of Things create the next BIM-BEM technological level, which lets us monitor our buildings' energy use and performance in real-time. Digital technology creates a complete building model that shows its current energy usage after gathering data from IoT sensors. This system lets owners watch energy patterns as they evolve. The dynamic nature of digital twins responds automatically to building usage habits, environmental shifts, and operational requirements, making energy model data stay true and up to date (Cheng et al. 2024, Daniels 2023).

Digital twins use IoT data to handle building energy planning across project stages through operations and repairs. The digital twin system receives HVAC performance data through IoT sensors to help operators examine performance problems and make energy changes. IoT sensors send real energy measurements to the digital twin AI for optimisation, which creates a repeatable system that improves energy performance (Skrzypczak et al. 2023).

The digital twin's predictive power enables facility managers to find problems earlier and maintain better building performance while making equipment last longer. Digital twins show how energy use changes over time so leaders can make decisions and support strategies that keep buildings environmentally friendly. Digital twins and IoT improvements will enlarge their contribution to BIM-BEM integration to track energy use and enhance sustainability performance better than ever before (Eames and Vinai 2024).

6.1.1 Addressing the Skills Gap in BIM-BEM Integration

The industry needs to tackle the skills gap that exists within BIM-BEM integration practices. The effective integration between Building Information Modelling (BIM) and Building Energy Modelling (BEM) requires a solution to the existing skills deficit within the industry. Here are targeted strategies to enhance education and professional development in this area:

6.2 Key Strategies for Skill Development

6.2.1 Curriculum Enhancement

Universities need to develop specialised coursework that teaches BIM and BEM applications for energy efficiency so that students gain appropriate knowledge. Students acquire practical skills through direct experience by completing projects that use standard software in the industry.

6.2.2 Professional Certification

Specialised training programmes for professionals who want to understand BIM-BEM integration better will create improved employment prospects. Workshops and online courses provided by our organisation will help industry professionals remain informed about contemporary technologies and practices.

Industry Collaboration: Educational content development through partnerships with technology firms ensures both current industry practices and practical relevance in curricula. Students, along with new professionals, receive practical training through BIM- and BEM-focused internships and apprenticeships at industry partner companies.

Mentorship Initiatives: New waves in BIM-BEM integration benefit from mentorship programmes that connect experienced professionals with new entrants to share essential knowledge about practical BIM-BEM challenges.

6.2.3 Online Learning Opportunities

By using digital networks, professionals can access training programmes for different skill levels through adaptable learning systems.

The successful execution of these educational programmes creates opportunities to close the skills gap that exists in BIM-BEM integration knowledge. Through educational advancement in curricula and professional certification programmes, together with industry mentorships and online learning access, the construction industry will train employees to use these advanced technologies competently. The planned approach enables the skilled growth of individual building experts while establishing more effective energy management practices across all buildings.

7. Conclusion

Combining Building Information Modelling (BIM) with Building Energy Modelling (BEM) leads us towards better energy-efficient designs that use resources responsibly while remaining affordable. By connecting BIM's digital data management and project collaboration with BEM's energy efficiency analysis, BIM-BEM integration helps teams find better ways to reduce building energy costs and environmental effects before the design phase ends (Cheng et al. 2024, Skrzypczak et al. 2023).

In Norway, a hospital project team used BIM-BEM integration to improve insulation and HVAC settings, which reduced annual energy costs by 30%, according to Ambrosini et al. (2023). Additionally, BIM-BEM integration creates a smooth workplace for design team members who can see real-time project updates across all roles. Joint access to energy goals during all project phases—from design to completion—helps projects meet LEED and BREEAM standards (Fonseca Arenas and Shafique 2023, Daniels 2023).

Advanced technologies, including artificial intelligence simulations and digital twins, help connect BIM-BEM systems by running automatic energy assessments and providing predictive data while allowing managers to view energy results at all times. The adoption of digital twin technology at an office building in Singapore produced real-time energy use assessments, saving 25% in energy costs per year (Eames and Vinai 2024).

7.1 Call to Action for Further Research and Standardization

To unlock the total value of BIM-BEM integration, researchers need to develop better data exchange methods and establish universal standards. Even though IFC and gbXML have helped advance progress, they have not solved all the problems yet. The latest research shows that data exchanges between Revit (BIM) and EnergyPlus (BEM) result in essential thermal data loss, which causes inaccurate energy simulation results (Cheng et al. 2024).

Standardisation progress needs work from buildingSMART and other groups to improve how construction data gets exchanged. Research into AI and digital twin technology will help BIM-BEM integration work better as it develops enhanced predictive tools to manage building energy use throughout its entire life cycle. Energy efficiency rose by 15%, according to data from the Oslo Living Lab's AI continuous energy monitoring system across various environmental settings. The construction industry must support BIM-BEM technology advancement to create a more sustainable sector (Skrzypczak et al. 2023).

References

"Designing with Climate: Bioclimatic Approach to Architectural Regionalism" - Focuses on integrating climatic conditions in building design for energy efficiency.

"Green Building and LEED Core Concepts Guide" - An introduction to green building principles and LEED certification processes.

"Sustainable Building Design: Principles and Practice" - A holistic view on sustainable practices in building design.

Keeping, M. and Shiers, D. (eds.) 2018. Sustainable Building Design: Principles and Practice. Wiley.

Olgyay, V. 1963. Design with Climate: Bioclimatic Approach to Architectural Regionalism. Princeton University Press.
U.S. Green Building Council. 2011. Green Building and LEED Core Concepts Guide (3rd ed.). USGBC.

Research Papers and Articles

"Innovations in Building Energy Modeling: Research and Development Opportunities for Emerging Technologies" 2023 - Explores new research avenues and technological advancements in building energy modeling Energy Efficiency and Renewable Energy
"Performance of Building Energy Modelers: A Review" 2022 - Analyzes the effectiveness and accuracy of various building energy modeling tools.
"Assessing the Impact of Building Energy Codes on Energy Consumption" 2023 - A study on how energy codes influence overall energy usage in buildings.
"Machine Learning Techniques in Building Energy Modeling" 2024 - Discusses the integration of machine learning in enhancing energy modeling accuracy.
"The Role of Building Energy Modeling in Achieving Net Zero" 2023 - Highlights the critical role of modeling in reaching net-zero energy targets.
"Building Energy Modeling and the Future of Energy Efficiency" 2022 - Reviews the current state and future trends in building energy modeling.
"Integration of Renewable Energy Sources in Building Design" 2023 - Examines strategies for incorporating renewable energy into building designs.
"Impact of Climate Change on Building Energy Demand" 2023 - Investigates how changing climate conditions affect energy needs in buildings.
"Simulation of Building Performance: The Future of Energy Efficient Design" 2022 - Discusses simulation techniques for assessing building performance.
"Energy Use in Commercial Buildings: Trends and Outlook" 2023 - An analysis of energy consumption patterns in commercial properties.
"Barriers to Energy Efficiency in Existing Buildings" 2023 - Identifies challenges in implementing energy-efficient upgrades in older buildings.
"Future Directions for Building Energy Research" 2024 - A forward-looking perspective on emerging research areas in building energy efficiency.
"Energy Modeling for Sustainable Design: Tools and Techniques" 2023 - A guide to various tools and methodologies for energy modeling in sustainable architecture.
"Evaluating the Economic Impacts of Energy Efficiency Investments" 2022 - Discusses the financial benefits of investing in energy-efficient building practices.
"Data-Driven Decision Making in Building Energy Management" 2023 - Examines how data analytics can improve energy management decisions in buildings.
"The Interplay of HVAC Systems and Energy Consumption in Buildings" 2022 - Looks at the relationship between HVAC systems and overall energy usage.
"Building Energy Modeling: Challenges and Future Research" 2023 - Reviews existing challenges and proposes future research directions in the field.
"Comparative Study of Building Energy Modeling Tools" 2024 - A comprehensive comparison of different energy modeling software tools and their applications.
Arenas, N. F. and Shafique, M. 2023. Recent progress on BIM-based sustainable buildings: State of the art review. Developments in the Built Environment 15: 100176.
Pan, Y., Zhu, M., Lv, Y., Yang, Y., Liang, Y., Yin, R. et al. 2023. Building energy simulation and its application for building performance optimization: A review of methods, tools, and case studies. Advances in Applied Energy 10: 100135.

Chapter 9

Evaluating the Integration of Industry 4.0 and Circular Economy Practices within the Manufacturing Sector

Abhishek Sahu,[1,*] *Saurabh Agrawal*[2] and *Neha Khandelwal*[3]

1. Introduction

India's biggest opportunity to drive economic growth and job creation in the coming decade lies in a manufacturing sector that is globally competitive. India is well-positioned to engage actively in global markets with expanding electricity capacity, long-term employment opportunities, and skill pathways for millions of people. This potential is strengthened by the nation's abundant raw materials, industrial expertise, and entrepreneurial spirit, which serve as foundational strengths. India is making significant progress towards Industry 4.0 (I4.0) through the execution of government initiatives, namely the National Manufacturing Policy, which proposes to raise the manufacturing sector's contribution to the country's GDP to 25% by the year 2025, and the PLI scheme for manufacturing, which was announced in 2022 with the

[1] Assistant Professor, School of Business, UPES, Dehradun, India.
[2] Associate Professor, Head of Department, Delhi School of Management, Delhi Technological University, Delhi-110042, India.
[3] Assistant Professor, Department of Finance, MIT Art Design and Technology University, Pune, India.
Emails: agrawals.iit@gmail.com, nehaalwar29@gmail.com
* Corresponding author: er.abhisheksahu2012@gmail.com

DOI: 10.1201/9781003475385-9

intention of bringing India's central manufacturing sector up to par with international manufacturing principles. I4.0 is focused on attaining smart manufacturing, technical podia, market responsiveness, smart products, and operational flexibility through increasing operational efficiency, reducing operational costs, and addressing the growing demand for personalised goods and services. As a result, Industry 4.0's main objective is to implement consumption-based or customised manufacturing at the inward and outward organisational levels in order to improve productivity, flexibility, innovation, profitability, and performance (Kamble et al. 2018). The other aim of I4.0 is to improve the connectivity between the various phases of the manufacturing process and consumer demands. Bag et al. (2020) identified the various technologies of I4.0, including Cyber-Physical Systems (CPS), Internet of Things (IoT) and Big Data (BD). These modern technologies have the potential to transform the present Indian manufacturing organisation.

In addition, the circular economy (CE) focuses on end-of-life perception by converting the linear economy model (take-make-use-dispose) into a circular formation (take-make-use-reuse). It is mainly centred on the three pillars of sustainable development, i.e., economic, environmental, and social. Moreover, blue economy, cradle-to-cradle, industrial ecology, and performance economy concepts are considered systematically by CE while designing new, sustainable business strategies (Sahu et al. 2022). In their work, Kristoffersen et al. (2020) explained that implementing the framework of the 3 Rs—reduce, reuse, and recycle—into the supply chain has many advantages for CE, such as preventing further environmental deterioration and gaining high levels of social capital, sustainability, and economic value. India's manufacturing sector has great potential in developing the economy as well as generating employment in global markets. These include an expanded electricity grid, long-term job creation, and skill-offering opportunities for millions of people (Palsodkar et al. 2024). India's availability of raw materials, industrial capacities, and enterprise culture form a strong base to engage in global value chains. Modernising and growing the manufacturing sector continues to be a focal area of growth for the country, and policies like the National Manufacturing Policy and the PLI scheme are laying the basis for an improvement in the GDP contribution of the manufacturing sector in India. Through such policies, the country seeks to raise the manufacturing industry to the global level and stimulate competition and development (Chhabra and Kr Singh 2024).

The execution of I4.0-based technology from the perspective of CE provides considerable potential for enhancing the position of the Indian manufacturing sector. It centres on improving manufacturing performance, driving down costs, and fulfilling consumer needs, as part of I4.0's emphasis on smart production, technical interfaces, and market dynamism. Technologies including IoT, CPS, and BD are revolutionising intermediate industries such as manufacturing by enabling real-time and timely decision-making (Kumar et al. 2021, Stentoft et al. 2021). This chapter explores how the move from a linear economy to a CE in the manufacturing industry is significant to the realisation of sustainable development goals (SDGs). When the main tenets of CE are integrated into Industry 4.0, it gives Indian manufacturers a platform to use resources efficiently, reduce waste, and enhance environmental and economic performance. In addition, this integration promotes a more efficient production cycle by increasing the flow of information within the value chain, thus making businesses accountable to their consumers as well as to sustainability goals (Piyathanavong et al. 2024).

Therefore, this study aims to integrate I4.0 drivers into the CE implementation. Prior research has concentrated on the basic design principles of I4.0 to obtain CE to expand production capabilities and improve product quality while maintaining a lower cost. Smart technologies could be important drivers of CE by providing real-time monitoring of the flow of goods, types of equipment, and materials. It enhances data accessibility, facilitating more efficient resource management and up-to-date decision-making at every stage of the product lifecycle. Furthermore, the enhanced information sharing along the entire value chain assists in the control and real-time regulating of processes in response to changing demand. The study focuses on investigating the incorporation of I4.0 technologies with CE practices to enhance sustainability in the Indian manufacturing sector. Based on this perspective, the following research questions are established in this study.

a) How do I4.0 drivers, such as IoT, CPS, and BD, influence the implementation of CE practices in India's manufacturing sector?
b) What is the impact of integrating I4.0 technologies on the economic, environmental, and social sustainability of the Indian manufacturing industry?

One of the key elements influencing the manufacturing of green products is consumer behaviour regarding the use of green products. Although extensive research has been conducted in this area, limited attention has been given to a detailed analysis of the causes of green consumption, particularly from the perspective of developing nations. Additionally, by operationalising and confirming the connection between green production and consumer aspirations in India, the present study aims to close the research gap. Further, this study has utilised the Partial Least Square Path Modelling (PLS-PM) approach of structural equation modelling (SEM) to determine the impact of IoT, BD, and CPS and their influences on the development of CE in India. As normalisation is not necessary for PLS-PM, it can be applied to small data sets. This approach has been utilised by several researchers. Smart PLS version 3.0 of the programme has been employed in this study because of its accuracy with a smaller sample size.

The study's main objectives are to:

a) To measure how the various I4.0 drivers help in CE implementation.
b) To determine how I4.0 drivers affect economic, environmental and social sustainability of the manufacturing industry.

The remaining part of this chapter is structured as follows: Section 2 deliberates the important theory and principles of I4.0 and CE. The research methodology of structural equation modelling, along with framework development and hypotheses, is described in Section 3. The results and discussion of the analysis are provided in

Section 4. Finally, conclusions, limitations, and research implications, along with future studies, are explained in Section 5.

2. Literature Review

The key drivers of I4.0 are CPS, IoT, and BD. BD is a technique used for processing a large amount of data or information, which encompasses both structured and unstructured data, and includes data collection, security, transport, storage, analysis, search, and visualisation (de Mattos Nascimento et al. 2024). In order to increase production, efficiency, and dependability, the IoT is a modern industrial network that combines intelligent, autonomous devices, cutting-edge predictive analytics, and seamless interaction between machines and humans. IoT offers real-time monitoring and sensing capabilities and quick data/information transmission capability, considerably facilitating remote manufacturing operations and effective stakeholder collaboration. IoT technology is utilised for data collection, analysis, and management, and object-oriented RFID systems are used in industrial systems as a control process (Skalli et al. 2024). Cloud manufacturing makes it possible for the manufacturing sector, which is a component of the smart, networked world, to connect manufacturing products, services, data, and people via the Internet. In the upcoming new era of industrial production, it is conceivable that I4.0 and cloud manufacturing are expected to unlock the great potential of the manufacturing sector. I4.0 can make extensive use of cloud technology to lower costs by putting systems online, improving system performance (such as enhanced flexibility and agility), and increasing data exchange across business borders. According to CPS, "end-to-end information communication technology-based integration" serves as a cornerstone of the fourth industrial revolution. CPS, short for "cyber-physical systems," refers to advanced technologies that seamlessly combine computational and physical functionalities while enabling interaction with humans through innovative modalities (Lu et al. 2024). Additionally, additive manufacturing is a method of generating three-dimensional objects by constructing them layer by layer. Each ultra-thin layer of material, either partially or fully melted, bonds to the previous layer. This process relies on computer-aided design (CAD) software to generate digital models, which are subsequently "sliced" into minute layers, allowing the object to be defined and constructed digitally. When a nozzle or print head precisely places material on the previous layer, this information directs its route. A bed of powdered material can selectively melt or partially melt under the influence of an electron or laser beam. A three-dimensional object is formed as materials solidify or undergo curing during the process.

According to Ellen MacArthur Foundation (2013), CE represents a paradigm shift in the usage of natural resources. The closed flow of materials, which reduces and minimises system inputs (raw materials, water, and energy) and undesirable outputs (waste and emissions), forms the basis of the CE. Moriguchi (2007) emphasised the need to preserve the value of goods, materials, and services by providing possible alternatives using Rs covered by the EC (CER) principles. This suggests switching to a circular economic system where "reduce, reuse, and recycle" are essential for maximising circularity. CERs incorporated into this model are environmental responsibility, social responsibility, and ethical governance practices (Gusmerotti et al. 2019, Sahu et al. 2023).

The focus is on minimizing input resource consumption, including raw materials, energy, and water, through smarter product utilization and production processes. Rejecting unnecessary items, rethinking their design (seeking multifunctional products), manufacturing them (simplifying and decreasing resource consumption), and using them more intensively, such as via sharing products, are all included in this concept (Kirchherr et al. 2017, Bhatia and Khandelwal 2024).

- *Reuse*: Extending a product's useful life directly is the objective of this R. When they perform their original purposes, it focuses on the reuse of goods or parts from the same type of product that was discarded by a previous user (Hussain and Malik 2020).
- *Recovery*: This R entails steps to enhance the useful life of products and the parts that make them up. These acts include repair (maintaining or fixing defective items), reconditioning (updating and modernising old products), remanufacturing (using components in the same type of products), and components reuse (using components in several types of products) (Jabbour et al. 2019).
- *Recycling*: This method seeks to put the material to good use. To replace the utilisation of natural resources, the method should concentrate on obtaining high-quality materials (Kirchherr et al. 2017).
- *Reduce emissions and waste*: This follows from the other Rs. Resource recovery ought to be the main concern. After the process is complete, CE demands a waste and emissions management strategy that aims to maximise resource recirculation while reducing environmental effects (Sahu et al. 2023).

In addition, I4.0 and CE have been considered cost-driven activities, and the performance's economic components have received the majority of attention. Rahman et al. (2022) observed that industries are implementing I4.0 and CE practices for financial reasons in a survey of the Indian manufacturing sector. Nowadays, manufacturing industries are also addressing environmental perspectives by contributing to I4.0 and CE (Gatell and Avella 2024). In the context of this study, we anticipate that developing economies adopt technology more quickly than developed economies because the former traditionally aspire to superior economic and social conditions. As a result, we suggest the hypotheses as follows:

H1: Cyber-Physical-Systems has a stronger favourable association with the CE in Indian manufacturing industry.

H2: Internet of Things has a stronger favourable association with the CE in Indian manufacturing industry.

H3: Big Data has a stronger favourable association with the CE in Indian manufacturing industry.

The following are the developed hypotheses.

H4. Environmental performance of I4.0 is positively correlated with the effectiveness of CE.

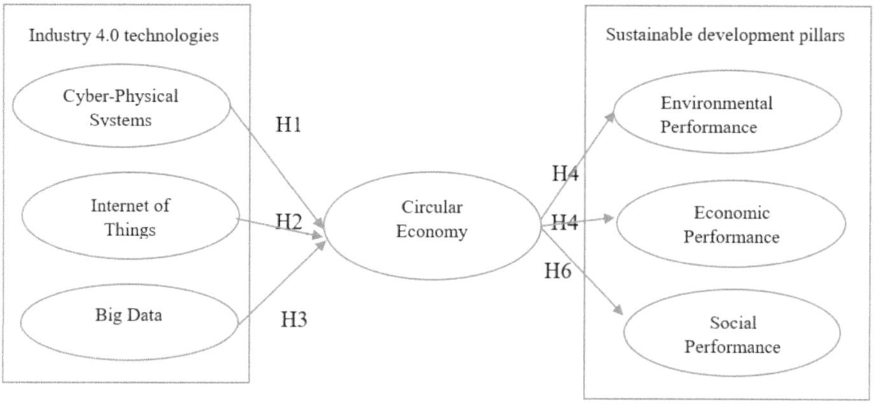

Figure 1. Model of hypothesised relationships. Source: Authors' composition.

H5. The economic performance of I4.0 is positively correlated with the effectiveness of CE.

H6. Social performance of I4.0 is positively correlated with the effectiveness of CE.

Figure 1 illustrates the model of hypothesised relationships. To verify and validate the hypotheses, the population framework for the study is limited to the Indian manufacturing industry. After assessing the data's fitness through the measurement model, the hypotheses are tested.

3. Research Methodology

The PLS-PM method of structural equation modelling, which considers the causal relationship among latent variables, was utilised to study relationship complexity. Due to its small sample size and ability to combine the benefits of multiple regression and factor analysis, PLS-PM was selected for the investigation. Researchers like Rahman et al. (2022) have employed PLS-PM in the field of logistics management, despite the fact that it is an older technique (2016). Furthermore, Singh (2019) applied structural equation modelling to measure disposition decisions to achieve triple bottom line aspects of sustainability for the Indian electronics industry. Other fields, including physics, mathematics, construction, and business and management, have also used it. The next section provides a thorough explanation of the research methodology.

3.1 Data Collection

A cross-sectional quantitative survey method is utilised for data collection from the perspective of Indian manufacturing industries. The present manufacturing industries have been selected because of the requirement for circularity performance, sustainable development objectives, and I4.0 capabilities. The data were circulated to 550 manufacturing industries through Google Forms that implemented I4.0-based technologies in their organisations. The study also selected industries that have already utilised CE concepts in their operations. Further, an additional 120 emails were sent to academia and researchers in this area. A total of 670 emails were circulated for conducting the survey. Out of 550 data samples, a total of 378 responses were received and 172 were incomplete; therefore, these incomplete responses were discarded. Hence, the response rate of the questionnaire is around 68.72%. Malhotra and Grover (1998) suggested that more than 20% of responses are recommended in operational management research.

3.2 Questionnaire Development

A questionnaire is established based on the hypotheses proposed for the study and is intended to collect data related to circularity decisions and the performance of I4.0 and CE from the viewpoint of the Indian manufacturing sector. In addition, five experts from academia and I4.0 practitioners from the Indian manufacturing industry reviewed the primary draft of the questionnaire. The experts examined the survey draft and provided comments on the content, readability, and representativeness of the questionnaire. The experts' opinions were incorporated, and the questionnaire was revised before distribution for data collection. Google Forms was utilised to establish the survey form. The survey form was sent through email, post, and industrial visits to the respondents, accompanied by a cover letter.

3.3 Hypotheses Testing Methodology

The acquired data was numerically coded in a Microsoft Excel spreadsheet before using the Smart PLS software for additional analysis. The missing numbers and inaccuracies in the data were visually checked for and modified where necessary. Structural equation modelling was applied to measure both measurement models and structural models. It is possible to analyse the measurement models using structural equation modelling inside the same analysis.

4. Results and Discussion

PLS-PM utilises both formative and reflective measurement model methodologies. The following criteria can be used to assess the reflective measurement models: convergent validity, discriminant validity, indicator reliability, and internal consistency reliability. The ability to accurately measure the same phenomenon repeatedly is demonstrated by a construct's reliability. The indicator's outer loading can be used to determine its dependability; it should be more than 0.7. However, to increase the composite reliability and average variance extracted (AVE), the indicator's loading between 0.4 and 0.7 can be removed. With the aid of Cronbach's alpha, which evaluates reliability based on the inter-correlation of the indicators, the internal consistency can be determined using composite reliability.

In addition, Cronbach's alpha value should range from 0.7 to 0.8. A value of less than 0.6 demonstrates a lack of reliability in internal consistency. As a result, in this study, the indicators whose outer loading was less than 0.5 have been eliminated in order to increase the construct's overall reliability. Convergent and discriminant validity can be used to evaluate validity. The AVE, whose value should be greater than 0.5, can be used to establish convergent validity, which describes the extent to which the indicators positively correlate with alternative indicators of the same construct.

Reflective measures can be used to analyse the concept of convergent validity in formative measurement models by comparing it to another measure of the constructs. The constructs should have a correlation of at least 0.7. Collinearity is a phenomenon where one predictor (indicator) may be accurately predicted from the other predictors to a significant extent. The variance inflation factor (VIF) can be used to determine whether the indicator is collinear. The VIF demonstrates the extent to which collinearity has raised the standard error. The VIF value should ideally be below 3; when it is between 3 and 5, collinearity becomes a concern. The indicator should be kept in the model if the outer loading value is greater than 0.5; otherwise, it should be eliminated. Both formative and reflective measurement models have been used in this investigation. From the data shown in Table 1, it is evident that the greatest VIF value is 1.747, which is considerably lower than the crucial value

Table 1. Results of the measurement model.

Construct	Indicators	Loadings	Collinearity (VIF)	AVE	Cronbach's Alpha	CR
Big Data (BD)	BD1	0.704	1.153	0.555	0.897	0.987
	BD2	0.844	1.379			
	BD3	0.675	1.241			
Circular Economy (CE)	CE1	0.610	1.099	0.576	0.723	0.896
	CE2	0.846	1.466			
	CE3	0.868	1.536			
Cyber-Physical-Systems (CPS)	CPS1	0.700	1.214	0.569	0.731	0.798
	CPS2	0.832	1.236			
	CPS3	0.725	1.282			
Economic Performance (ECP)	ECP1	0.796	1.487	0.701	0.787	0.876
	ECP2	0.867	1.842			
	ECP3	0.848	1.747			
Environmental Performance (ENP)	ENP1	0.698	1.402	0.616	0.799	0.827
	ENP2	0.825	1.587			
	ENP3	0.825	1.279			
Internet of Things (IoT)	IoT1	0.731	1.366	0.600	0.774	0.818
	IoT2	0.784	1.372			
	IoT3	0.806	1.235			
Social Performance (SP)	SP1	0.631	1.145	0.519	0.847	0.961
	SP2	0.873	1.228			
	SP3	0.631	1.126			

Source: Authors' composition

of 5 and below the threshold value at which collinearity becomes a problem for PLS-PM estimation. The formative indicators' relevance and importance are evaluated by their outer weight. The outcome of multiple regression using the construct as a dependent variable and an indicator of the independent factors is the outer weight. The construct's respective relative contribution is represented by the outer weight. The collinearity of the indicators, through the VIF value of the formative indicator, has also been examined in this study.

Table 2 displays the discriminant validity measurements for a reflective construct using the Fornell-Larcker criterion. A convergent validity test has been used to examine the relationship between one measure and other measures of the same concept that employ different indicators. The construct must be original and cannot include phenomena that are covered by other constructs. The validity of discrimination can prove this. Cross-loading techniques, the Heterotrait-Monotrait Ratio (HTMT), and the Fornell and Larcker approach can all be used to estimate discriminant validity. The loading of constructs is tested in a cross-loading procedure and should be higher than their cross-loadings. To assess discriminant validity, Fornell and Larcker (1981) recommended that the "square root" of the AVE of each latent variable should be greater than the correlations among the latent variables. While the value of HTMT remains constant in cross-loading, it can change when additional constructs are added to the model. In order to achieve favourable outcomes, the HTMT value needs to be smaller than 0.85.

The highest potential outer weight is 1/n, where n is the number of indicators if there is no correlation between the indicators. The indicator would typically be retained despite an insignificant outer weight when outer loading is more than 0.50. The bootstrapping procedure verifies the outer weight's significance. All of the formative indicators in this study are significant at the 5% level, as evidenced by the outer weight/loading in Table 1 and the 'p' and 't' values in Table 3. Further, results obtained from the Smart PLS version 3.0 software are presented in Figure 2.

Table 2. Result of Discriminant Validity Test (Fornell Larker Criterion).

	BD	CE	CPS	ECP	ENP	IOT	SP
BD	**0.745**						
CE	0.460	**0.759**					
CPS	0.493	0.696	**0.754**				
ECP	0.443	0.724	0.737	**0.838**			
ENP	0.410	0.672	0.616	0.628	**0.785**		
IOT	0.497	0.579	0.644	0.596	0.556	**0.774**	
SP	0.398	0.654	0.668	0.673	0.620	0.530	**0.720**

Source: Authors' composition

Table 3. Hypothesis test results.

Paths	Hypothesis	Path coefficient/R^2	p Value	Results
Effects on CE		$R^2 = 0.522$		
By CPS	H1	0.519	0.000	Supported
By IOT	H2	0.190	0.005	Supported
By BD	H3	0.110	0.051	Not Supported
Effects on ENP		$R^2 = 0.451$		
By CE	H4	0.672	0.000	Supported
Effects on ECP		$R^2 = 0.524$		
By CE	H5	0.724	0.000	Supported
Effects on SP		$R^2 = 0.428$		
By CE	H6	0.654	0.000	Supported

Source: Authors' composition

The coefficient of determination (R^2), effect size (f^2), and statistical significance of the structural path coefficients can all be used to evaluate the structural model. The model's prediction precision is estimated by R^2. In marketing research, R^2 values of 0.75, 0.5, and 0.2 are considered substantial, moderate, and weak, respectively. The f^2 effect size estimates the impact of omitted constructs as a result of changing the R^2 value.

The path coefficients and other model fit indices for this model are shown in Table 3. The statistical analysis provides the coefficient of influence and the p-values. Finally, all paths are positive in the model and statistically significant except for BD. The key relationship is then explored more deeply by the model. CPS (H1, $\beta = 0.519$, p-value = 0.000) and Internet of Things (H2, $\beta = 0.190$,

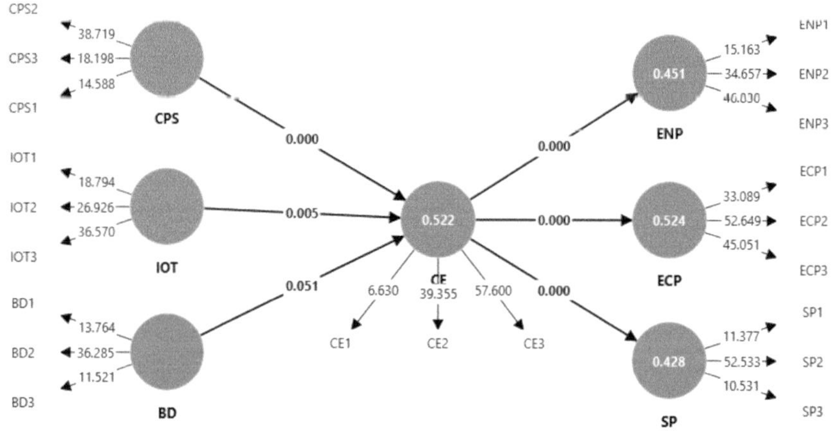

Figure 2. Structure equation model for implementation of I4.0 drivers in CE. Source: Authors' composition.

p-value = 0.005) results indicate a positive and significant impact on CE. Nevertheless, the impact of BD on the CE is positive with no statistical significance (H3, $\beta = 0.110$, p-value = 0.051). Furthermore, the CE partially mediates the environmental (H4, $\beta = 0.672$, p-value = 0.000), economic (H5, $\beta = 0.724$, p-value = 0.000), and social performance (H6, $\beta = 0.654$, p-value = 0.000). The fit of the CE can be explained by the CPS, IoT, and BD to 52.2% ($R^2 = 0.522$). The "CE" explains 45.1% ($R^2 = 0.451$) of "Environmental Performance", 52.4% ($R^2 = 0.524$) of "Economic Performance" and 42.8% ($R^2 = 0.428$) of "Social Performance", showing the model prediction power.

The findings of this study deliver a strong indication supporting the role of CPS and IoT in driving the CE, which is consistent with earlier research that has investigated the impact of I4.0 technologies on sustainable operations. Previous researchers have emphasised that CPS and IoT contribute significantly to the CE by facilitating better data management, process optimisation, and waste reduction. For example, studies by Lei et al. (2023) discuss how CPS can enable more efficient production systems and better monitoring of resource usage, which aligns with our findings that these technologies significantly influence the CE.

In contrast, the lack of statistical significance of BD on the CE in this study presents a deviation from prior research. Earlier studies have suggested that BD could play a major role in enhancing CE practices by improving decision-making, enabling predictive maintenance, and optimising resource flows (Schneikart et al. 2024). The lack of a significant relationship in this study could reflect challenges related to BD implementation, such as data integration issues or a lower level of maturity in BD adoption in certain industries—an area that requires further exploration in future research.

The study also found that the CE partially mediates the impact of I4.0 technologies on environmental, economic, and social performance, which is supported by the work of Sahoo and Jakhar (2024). These authors argue that the implementation of CE principles leads to positive environmental outcomes by reducing waste, while also driving economic benefits through increased efficiency and reduced costs. Similarly, CE practices are known to bring social benefits by encouraging innovation, creating jobs, and fostering a better quality of life (Suchek et al. 2024).

Additionally, the explanatory power of the CE model in this study, where it describes a substantial amount of variance in environmental, economic, and social performance, reinforces the findings of Ghaithan et al. (2023), who also identified CE strategies as key enablers of sustainability. This research encompasses the existing literature by empirically validating the role of CPS, IoT, and BD in CE practices and measuring their subsequent influence on sustainable performance metrics. Overall, the findings suggest that the integration of I4.0 technologies with CE principles has the potential to substantially enhance sustainability outcomes, confirming the importance of technological innovation in achieving long-term environmental, economic, and social objectives (Kusi-Sarpong et al. 2023).

The implications of this research provide valuable insights for businesses and legislators motivated to foster sustainability. As industries move towards more sustainable practices, the integration of modern technologies including CPS and IoT can play a critical role in advancing CE principles. By leveraging these technologies, businesses can optimise resource management, reduce waste, and increase operational efficiency. This shift towards smarter, data-driven operations not only helps reduce environmental impact but also unlocks a novel path for cost savings and value creation. Therefore, organisations should prioritise investments in these technologies to stay competitive in a sustainability-driven market, ensuring that their operations align with CE objectives. However, addressing the challenges around the effective implementation of BD is necessary to fully realise its potential in driving sustainable outcomes.

From a policy perspective, supporting the transition towards a CE requires creating an environment that encourages the adoption of sustainable practices. Policymakers should provide incentives for businesses to integrate CE models, including offering tax breaks or funding for research and development in sustainable technologies. Additionally, creating standards for data sharing and improving collaboration between industries will be essential to maximising the potential of IoT and BD for CE initiatives. Governments can also implement regulations that encourage businesses to consider the environmental, economic, and social implications of their procedures. By fostering a supportive policy environment, governments can help reduce barriers to the adoption of CE principles and technology, facilitating broader societal and environmental benefits.

5. Conclusion

The findings of the study demonstrated that I4.0 implementation in CE is essential to overcoming the many challenges that Indian manufacturing industries are currently facing. In particular, all four of I4.0's technologies taken into consideration play a significant role in enhancing circularity and overall performance in the Indian manufacturing industry. Regardless of the type of CE, the implementation of the attested technology has the potential to improve operational efficiency through real-time product traceability and reliability. I4.0 drivers such as CPS, IoT, BD, and CC also demonstrated that there is a beneficial relationship between them and the service industry, which eventually has a favourable impact on the overall performance of the manufacturing industry in India. Therefore, in order to implement a thorough and comprehensive strategic plan that addresses all the interrelated parts of I4.0 drivers, the manufacturing industries of India need to implement these latest technologies of I4.0. As was previously mentioned, I4.0 and CE work in harmony with the UN's SDGs to safeguard the environment by encouraging sustainable consumption and production. Therefore, this study aimed to explore how technological tools have affected the manufacturing industries in developing economies like India.

5.1 Research Implication

Findings demonstrate that industries in both developing and developed countries must concentrate on various areas of the technical instruments within I4.0 and CE. As a result, initiatives for achieving the UNSDG of sustainable production and consumption can be more unified and successful. This study has management implications because it focused on business performance using the most recent technologies while providing insight into the manufacturing industry. This survey may help managers and industrialists understand how I4.0 is improving the performance of the manufacturing industry as a whole. The results showed that, regardless of the various economies, I4.0 drivers can have a substantial impact. Consequently, industries should take the necessary steps to adopt the most recent technology in both short- and long-term strategic planning. Businesses, especially those in emerging economies, which have lower technology representation and subpar performance, can benefit from this study as it serves as an example of how numerous I4.0 drivers can be put into practice. Decision-makers can utilize the findings of study to raise the manufacturing industry's operational effectiveness and performance while resolving issues with technology.

References

Agrawal, S. and Singh, R. K. 2019. Analyzing disposition decisions for sustainable reverse logistics: Triple Bottom Line approach. Resources, Conservation and Recycling 150: 104448.

Bag, S., Yadav, G., Wood, L. C., Dhamija, P. and Joshi, S. 2020. Industry 4.0 and the circular economy: Resource melioration in logistics. Resource Policy 68: 101776.

Bhatia, A. and Khandelwal, N. 2024. Adapting to the Pandemic: The influence of digital media on the indian automobile sector. Journal of Informatics Education and Research 04(01): 1526–4726.

Chhabra, D. and Kr Singh, R. 2024. Analyzing barriers to green logistics in context of Circular Economy and Industry 4.0 in the Indian manufacturing industry. International Journal of Logistics Research and Applications 27(11): 1939–1952.

de Mattos Nascimento, D. L., de Oliveira-Dias, D., Moyano-Fuentes, J., Maqueira Marín, J. M. and Garza-Reyes, J. A. 2024. Interrelationships between circular economy and Industry 4.0: A research agenda for sustainable supply chains. Business Strategy and the Environment 33(2): 575–596.

Ellen MacArthur Foundation. 2013. Towards the Circular Economy, Economic and Business Rationale for an Accelerated Transition, Ellen MacArthur Foundation, Cowes.

Fornell, C. and Larcker, D. F. 1981. Evaluating structural equation models with unobservable variables and measurement error. Journal of Marketing Research 18(1): 39–50.

Gadekar, R., Sarkar, B. and Gadekar, A. 2024. Model development for assessing inhibitors impacting Industry 4.0 implementation in Indian manufacturing industries: an integrated ISM-Fuzzy MICMAC approach. International Journal of System Assurance Engineering and Management 15(2): 646–671.

Gatell, I. S. and Avella, L. 2024. Impact of Industry 4.0 and circular economy on lean culture and leadership: Assessing digital green lean as a new concept. European Research on Management and Business Economics 30(1): 100232.

Ghaithan, A. M., Alshammakhi, Y., Mohammed, A. and Mazher, K. M. 2023. Integrated impact of circular economy, industry 4.0, and lean manufacturing on sustainability performance of manufacturing firms. International Journal of Environmental Research and Public Health 20(6): 5119.

Gusmerotti, N. M., Testa, F., Corsini, F., Pretner, G. and Iraldo, F. 2019. Drivers and approaches to the circular economy in manufacturing firms. Journal of Cleaner Production 230: 314–327.

Hussain, M. and Malik, M. 2020. Organizational enablers for circular economy in the context of sustainable supply chain management. Journal of Cleaner Production 256: 120375.

Jabbour, C. J. C., de Sousa Jabbour, A. B. L., Sarkis, J. and Godinho Filho, M. 2019. Unlocking the circular economy through new business models based on large-scale data: an integrative framework and research agenda. Technological Forecasting and Social Change 144: 546–552.

Kamble, S. S., Gunasekaran, A. and Sharma, R. 2018. Analysis of the driving and dependence power of barriers to adopt industry 4.0 in Indian manufacturing industry. Computers in Industry 101: 107–119.

Kirchherr, J., Reike, D. and Hekkert, M. 2017. Conceptualizing the circular economy: An analysis of 114 definitions. Resources, Conservation and Recycling 127: 221–232.

Kristoffersen, E., Blomsma, F., Mikalef, P. and Li, J. 2020. The smart circular economy: A digital-enabled circular strategies framework for manufacturing companies. Journal of Business Research 120: 241–261.

Kumar, P., Singh, R. K. and Kumar, V. 2021. Managing supply chains for sustainable operations in the era of industry 4.0 and circular economy: Analysis of barriers. Resources, Conservation and Recycling 164: 105215.

Kusi-Sarpong, S., Gupta, H., Khan, S. A., Chiappetta Jabbour, C. J., Rehman, S. T. and Kusi-Sarpong, H. 2023. Sustainable supplier selection based on industry 4.0 initiatives within the context of circular economy implementation in supply chain operations. Production Planning & Control 34(10): 999–1019.

Lei, Z., Cai, S., Cui, L., Wu, L. and Liu, Y. 2023. How do different Industry 4.0 technologies support certain Circular Economy practices?. Industrial Management and Data Systems 123(4): 1220–1251.

Lu, H., Zhao, G. and Liu, S. 2024. Integrating circular economy and Industry 4.0 for sustainable supply chain management: A dynamic capability view. Production Planning & Control 35(2): 170–186.

Malhotra, M. K. and Grover, V. 1998. An assessment of survey research in POM: from constructs to theory. Journal of Operations Management 16(4): 407–425.

Moriguchi, Y. 2007. Material flow indicators to measure progress toward a sound material-cycle society. Journal of Material Cycles and Waste Management 9: 112–120.

Palsodkar, M., Yadav, G. and Nagare, M. R. 2024. Integrating Industry 4.0 and agile new product development practices to evaluate the penetration of sustainable development goals in manufacturing industries. Journal of Engineering, Design and Technology 22(4): 1351–1392.

Piyathanavong, V., Huynh, V. N., Karnjana, J. and Olapiriyakul, S. 2024. Role of project management on Sustainable Supply Chain development through Industry 4.0 technologies and Circular Economy during the COVID-19 pandemic: A multiple case study of Thai metals industry. Operations Management Research 17(1): 13–37.

Rahman, M., Kamal, M. M., Aydin, E. and Haque, A. U. 2022. Impact of Industry 4.0 drivers on the performance of the service sector: comparative study of cargo logistic firms in developed and developing regions. Production Planning & Control 33(2-3): 228–243.

Sahoo, S. and Jakhar, S. K. 2024. Industry 4.0 deployment for circular economy performance—Understanding the role of green procurement and remanufacturing activities. Business Strategy and the Environment 33(2): 1144–1160.

Sahu, A., Agrawal, S. and Kumar, G. 2022. Integrating Industry 4.0 and circular economy: a review. Journal of Enterprise Information Management 35(3): 885–917.

Sahu, A., Khandelwal, N. and Bhatia, A. 2023. Critical success factors of Industry 4.0 and circular economy to achieve green economy and sustainable development goals: a dematel approach. Korea Review of International Studies 16(09): 1226 474.

Schneikart, G., Mayrhofer, W., Löffler, C. and Frysak, J. 2024. A roadmap towards circular economies in pharma logistics based on returnable transport items enhanced with Industry 4.0 technologies. Resources, Conservation and Recycling 206. 107613.

Skalli, D., Charkaoui, A., Cherrafi, A., Garza-Reyes, J. A., Antony, J. and Shokri, A. 2024. Analyzing the integrated effect of circular economy, Lean Six Sigma, and Industry 4.0 on sustainable manufacturing performance from a practice-based view perspective. Business Strategy and the Environment 33(2): 1208–1226.

Stentoft, J., Adsbøll Wickstrøm, K., Philipsen, K. and Haug, A. 2021. Drivers and barriers for Industry 4.0 readiness and practice: empirical evidence from small and medium-sized manufacturers. Production Planning & Control l32(10): 811–828.

Suchek, N., Ferreira, J. J. and Fernandes, P. O. 2024. Industry 4.0 and global value chains: what implications for circular economy in SME?. Management Decision 62(9): 2915–2936.

Index

R

S

T

U

W

For Product Safety Concerns and Information please contact our
EU representative GPSR@taylorandfrancis.com Taylor & Francis
Verlag GmbH, Kaufingerstraße 24, 80331 München, Germany